BUILDING SIMPLE

MODEL STEAM ENGINES

Other books by the same author

Drills, Taps and Dies
Hardening, Tempering and Heat Treatment
Milling Operations in the Lathe
Model Engineer's Handbook
Soldering and Brazing
Spring Design and Manufacture
Workholding in the Lathe
Workshop Drawing

Writing as T. D. Walshaw

The I.S.O. System of Units
Ornamental Turning

TUBAL CAIN

BUILDING SIMPLE MODEL STEAM ENGINES

SPECIAL INTEREST MODEL BOOKS

Special Interest Model Books Ltd.
P.O. Box 327, Poole, Dorset BH15 2RG

First published by Model and Allied Publications 1981
Reissued by Argus Books 1993
Reprinted by Nexus Special Interests 1997, 1999, 2003

This edition published by Special Interest Model Books Ltd, 2004

www.specialinterestmodelbooks.co.uk

ISBN 1 85486 104 2

Printed and bound in Great Britain by
Biddles Ltd, *www.biddles.co.uk*

CONTENTS

INTRODUCTION

The Author with his first steam engine about 1920.

There is a fascination about the simple oscillating steam engine which attracts even the builders of true-scale, exact-to-prototype quadruple expansion marine engines. It may be the sheer simplicity of the mechanism, it may be memories of childhood days when "Father Christmas" put one in the stocking, or it may be just the fun of seeing the "works" work – or not seeing them as the outline of the connecting rod disappears into a blur. This little book describes the making of four such models.

I have built all four during the last few years, and all save the last have been featured in articles in the "Model Engineer" and elsewhere. In putting them into book form I have had two objects in mind. First, to provide the novice with designs and methods of construction which will enable him to bring his efforts to a successful conclusion. To this end I have not refrained from repeating instructions – nothing is more annoying than the frequent advice to refer back to a page in a previous chapter! In some cases different methods have been used for similar components in different engines; this will, I hope, point out the seldom realised truth that "there is no one correct solution to any engineering problem"! It also provides the constructor with the pleasures of variety! If more experienced readers find the detail too obtrusive I would urge them to remember that there was a time when they, too, were novices.

The second objective has been to offer to *all* Model Makers, beginners or Championship Cup Winners, a series of designs which will provide pleasure to children, and, perhaps, to start these youngsters on the model-making road. To win the Duke of Edinburgh's Award at Wembley is a great achievement, open only to the best of us; but I very much doubt if such an award can give as great a pleasure as the sight of a young child unwrapping "Uncle George's" (Or Auntie Cherry's) present of a little steam engine, and his delight at setting it to work.

I hope this book will meet these objectives. I can say that the engines described have given me pleasure in the making, and enjoyment to the youngsters who now own some of them.

Tubal Cain. November, 1980

Chapter One
WAYS AND MEANS

This chapter deals with a few special methods which may be needed in the making of the engines. I have, in other chapters, described those I used myself when making the models and have gone into them in some detail; I do not propose to cover them now. However, there are a few areas which warrant some general comment and it is the purpose of this chapter to anticipate any question that may arise.

Materials.

All the materials mentioned either on the drawings or in the description are readily available from a number of firms, most of whom advertise in the "Model Engineer" and similar magazines. However, constructors often wish to use materials already on their shelves but are doubtful whether they are suitable.

Boilers, at the low pressures used in these engines, can be made of brass if desired. The advantage is that it takes a better colour when polished. The disadvantages are, first, that after a long period of time the material may deteriorate – after some 60 years or so it may go brittle. This is a long time, but it is surprising how many generations may enjoy a well-made model! Second, brass has a lower melting point than copper and there is a slight risk of the metal running during the brazing operation if overheated. The melting point – about 890°C – is well above that of the usual silver-brazing alloy (silver solder) at around 630°C, but an accidental overheating may cause trouble. My advice to those brazing for the first time is to use copper – others may take their choice.

Pistons and Cylinders. The general rule is that rubbing surfaces should be of dissimilar metals. A brass piston in a brass cylinder will usually work, but the wear may be rapid, especially if there is a shortage of oil. The ideal combination is stainless steel in brass. Drawn gunmetal or phosphor bronze pistons in brass cylinders are satisfactory, and *cast* gunmetal will run in cast gunmetal – this combination is used in the crane. Some nickel-copper alloys (German Silver) work well with brass, and a german silver cylinder has the attraction of polishing *like* silver.

Steels. Where stainless steel is specified for a working part (e.g. a crankshaft) the object is only to avoid rusting. Mild steel will do instead apart from this. Silver steel – which is a high carbon steel – is, where called for, only because it can be obtained ground accurately to size and saves machining. There is seldom any need for the greater strength and hardness, and ordinary stock, or evcn ground stainless can be used instead. In some cases – e.g. the pivot pins for oscillating cylinders – a piece of hard brass wire or rod can be used; quite strong enough, and preferable to mild steel again because it will not rust. Either copper or brass tube can be used for steam pipes.

For such things as frames and brackets there is little to choose between brass or steel apart from such factors as cost or ease of bending. I use "Ternplate" – lead coated steel – quite a lot, as it is fairly easy to bend and solders easily. It is, in many ways, preferable to tinplate. I don't recommend the use of galvanised or zinc-coated steel, as the coating will react with any adjacent copper or brass and cause corrosion. The same applies to aluminium.

Bearing these notes in mind, the constructor has a pretty wide choice of materials he can use. Indeed, the engine described in Ch. III was made entirely of odds and ends I had lying about, and I didn't search my bit-box very thoroughly at that.

Dimensions.

The drawings show the engines as I made them. Sized thus they should work satisfactorily. But subject to one important condition, none is critical. That condition is that you make a new drawing incorporating the proposed changes. Unless this is done there is a considerable risk that parts other than those altered will not fit; flywheels may foul the boiler, gears may foul axles, or pipes may be too large to go into the space available. Enlargement of boiler size will give longer steaming times; but you must also enlarge the lamp. Reduction in boiler size is not recommended. One gauge thicker or thinner for the boiler shell will not hurt, and metric equivalent sizes need cause no concern.

Small changes in cylinder bore will have little effect, though it should be borne in mind that enlargement from (say) $\frac{1}{4}''$ bore to 5/16" will demand nearly 60% more steam! Changes in *stroke* should be avoided, as this means a wholesale revision both of port spacing and port size. These latter dimensions are very carefully worked out to give the desired performance, and should not be altered unless you are capable of doing the necessary geometric constructions. Having said that, some fun can be had by experimenting with port sizes, going up or down one drill size at a time, and if an engine runs sluggishly this may be worth doing. (Though it would be wise to check the fit of the piston first). It can happen that a drill will wander off course when drilling the holes in the stationary port block, and a change in the size of the ports may effect an improvement. The twin-port arrangement of the first engine is designed to avoid this problem.

Otherwise, it doesn't matter a puff of steam whether a crankshaft is $\frac{1}{8}''$ or 5/32" diameter – many commercial engines of this style had nothing but bent wire here – and bearing clearances are by no means critical either.

Finally, – lampwicks. It is by no means easy to get these nowadays, bottled gas having replaced the old paraffin lamps. Shredded asbestos string is a good substitute, and has the advantage that it doesn't char – though I don't think it gives as good a flame. Candlewick can be obtained, both from novelty shops where it is sold for making candles, and from drapers; I won't, however, suggest that you snip off the border of the candlewick bedspread!

Manufacturing methods.

Boilermaking. There is an unjustified apprehension on the part of many about "brazing". This probably springs from the days when we had to use true brazing spelter, needing high temperatures and much rubbing about with pointed iron wires to get the stuff to run. However, the advent of "Silver Solder" has changed all that; the silver in the alloy both reduces the temperature needed and increases the fluidity of the melt. In fact, I find it much easier to make a good job of silver brazing than I do with soft (lead) soldering. It is much easier to control, and, of course, much stronger. I have gone into the procedure in sufficient detail in the construction chapters to enable the rawest novice to achieve success, and will only say at this stage "Practice makes perfect" – try your hand on a few practice exercises first; you will very soon get the hang of it. For novices, "Easyflo" No. 1 is recommended, but the No. 2 alloy is a bit cheaper.

Turning Problems. What to do if you have no lathe? Well, both of the first two engines can be made without one if need be. But I must say that you need reasonable drilling facilities; I would expect you to have a fair bit of difficulty drilling even a No. 53 hole with a hand drill! The ordinary domestic electric drill will serve provided you have a *good* drill stand for it, one which gives you a sensitive feed and which will drill square to the table.

There are only two components on the first two engines which apparently require a lathe; the cylinder-piston assembly, and the flywheel. Let us look at these in turn. The piston rod is usually screwed into the piston, the hole being drilled in the lathe. In fact, it doesn't matter much if this hole isn't in the dead centre of the piston body – the important requirement is that the hole be in parallel line to the piston axis. This requires no more than care in the drilling – achievable even with a hand-drill.

The piston must fit the cylinder. There are two ways of doing this without a lathe. The first is illustrated in Fig. 3–1. We use a piece of material for the piston which is a slack fit in the cylinder, and make it steam tight with soft packing. The thing should fit reasonably well, of course, as the piston does enjoy another office – that of causing the cylinder to oscillate, so that it needs to act as a reasonable sliding bearing. However, given that, all that is necessary is to file a groove round the piston, keeping it as parallel as you can in width and depth. This is then packed with some soft material impregnated with lubricant; candlewick soaked in melted tallow is almost ideal. (Tallow, not ordinary candlewax; plumbers use it) The packing must not be very tight; just enough to make a seal..

The second method is almost the reverse. Choose a piece of rod which is a *tight* fit, or perhaps even too tight to fit, the cylinder. If oversize, spin it in your drill and reduce it with fine emery cloth; Tack the emery to a flat stick, which use like a file on the rotating workpiece; this will avoid getting it barrel shaped. When it is a tight fit, so that it can be slid right down the bore, but needs some force to do so, reduce it further by lapping. First, spin a piece of ordinary dowel in your drill and reduce it with sandpaper to a slide fit to the cylinder. Make sure no grains of sand are embedded, then anoint this with dried-up Brasso and a little oil, or really fine "flour emery" if you have any, and (with the dowel in the drill) apply it to the inside of the cylinder. If your drill has two speeds, use the slower – or use a hand drill. Reciprocate the cylinder up and down, so that the dowel just doesn't emerge completely each stroke. Don't go on too long – a minute should be enough. Then try the piston – it may fit as it is; it should be an easy slide fit. If it is a bit tight still, file the three oil grooves (use a 3-cornered needle file) anoint the piston with the the oil-Brasso mixture and lap with this; use the hand drill, though; a power drill is too fast. Aim at one oscillation every two rotations of the work. Renew the Brasso from time to time, cleaning all the old stuff off. This may take time, could be half an hour or so, but it is effective and will, in fact, give you a better fit to the piston than will turning it.

If you are fortunate to find a piece of piston material already a good fit, file the oil grooves and then lap off the burrs in the same way, but I have only once found a piece like this. (In fact, when building the little horizontal engine)

Flywheel This is a bit more difficult, but it is worth remembering that most of the commercial engines of 50 years ago simply had a cast lead wheel, drilled and pressed onto the shaft – itself a piece of thin wire! So, don't be alarmed – it is quite a practicable job.

It is possible to saw off the end of a bar by hand and still keep the disc of uniform thickness, but not easy, and it is hard work. But you can obtain from a number of suppliers "blanks", i.e. discs of almost any diameter and thickness at quite reasonable cost, in both brass and steel, the latter being the cheaper. (An aluminium flywheel is no good, I'm afraid!) Start with one of these. First file out the saw marks, if any, on the faces (some are pressed out of thick sheet) and if the thickness isn't quite the same all over, try and correct that, too. Don't polish the surface, rather roughen it at this stage using coarse emery.

Now, to establish the centre, set your dividers, or a pair of jenny calipers, at very nearly the radius of the disc. From various points round the circumference – at least three and preferably five – scribe arcs across the centre. These will meet if your setting is dead right, otherwise will enclose a small area within which the centre lies. Repeat the process if this is more than say 1/16" across. Then, using an eyeglass, set a small centre-pop in the middle of this area. Examine this, and if it seems satisfactory, deepen the centre. At this centre, drill through one size smaller than the size of your shaft; No. 31 for 1/8", No. 23 for 5/32", etc. Take very great care though to get the drill dead square to the surface.

Take a piece of shaft-size steel rod and spin in your electric drill, reducing the diameter with emery cloth till the rod is a tight fit in the hole. If by chance it is slack to start with, mount it with a spot of Loctite nut-lock. Hold the shaft in a hand-drill gripped in the vice and rotate. You will see if it shows any signs of wobble. If it doesn't, there is your flywheel! All you have to do is to enlarge the hole to suit the shaft. If the wobble is *very* small, it may still serve. But if there is perceptible wag, hold a pencil against the rim with one hand and rotate slowly with the other. File out the pencil marks, using a fine file wider than the rim. Repeat, and repeat, till the pencil marks all round. Obviously you should file off more in the centre of the pencil stripe than at the ends, but given time you should be able to get the wheel tolerably round.

If there is appreciable side-wobble, do the same here – and you will appreciate that you must take a bit off both sides, on opposite diameters in this case. Wobble on the centre third of the area won't matter. Again, this is just a case of taking time – there is no difficulty as such.

This method gives you no thrust face to bear on the framework so just use a washer. It doesn't give you a little boss to serve as a pulley either. So use a Meccano one – in fact, if you use 5/32" dia crankshaft for both the engines this will fit the Meccano pulley – all you have to do is to screw the end for the crankweb 5/32" x 40 tpi or whatever 5/32" tap and die you may have. Indeed, you can, these days, attach the crankweb with Loctite retaining compound. You will have no pretty recesses in the side of the wheel either. So, decorate them with paint instead, especially if your wheel is steel.

Mentioning Meccano, you can, of course, use a medium pulley from your set and load it up with lead rings, making an even more effective flywheel due to the greater mass. I'm afraid, however, that the Meccano flywheel isn't really heavy enough, being no more than two tinplate presssings joined together face to back.

Crankdisc. On the vertical engine this has a 5/16" dia boss. You can either make the disc that much thicker, or simply drill a piece of 5/16" brass to make a spacing washer. The attachment of the crank axle to the thinner disc will be quite strong enough. The main point about these discs, though, is to take care that the two holes – for axle and crankpin – are drilled square to the disc and parallel to each other.

Safety Valve It would be difficult to make these without a lathe, though it has been done. However, such fittings are available commercially, the only precaution you must take being to see that they are not too large – those sold for 3½" or 5" gauge locomotives just would not fit the boiler top. Those sold for 2½" gauge locos will be a tight fit to the vertical engine, but should do for the horizontal, though a bit overlarge in appearance. Gauge 1 or gauge 0 will serve well. Don't forget to order a bush with the valve, and ask for one set at 30lb. sq. in. Incidentally, if you can obtain a "spare part" valve for a "Mamod" or "Meccano" steam engine, this would be ideal. There would be no bush with it, of course, but this can be managed by hand filing if need be. At a pinch there is no reason why you should not use a simple piece of rod with the hole drilled through – the only reason for the shoulder on the bush is to make it easier to locate during the brazing operation.

Other Work. In the engine descriptions the lathe has been employed for many operations – trimming the ends of the boiler barrel, for example – solely because this is the convenient way of doing it. A file is equally effective, but needs the fairly frequent application of the try-square and takes longer, that is all. It is a regrettable fact that nowadays many jobs are dismissed as "impossible" because some special machine or other is not available. In reality, almost all can be carried to a successful conclusion with no more than a saw, hammer and chisel, file, and drill. I have seen a model traction engine towing a healthy load of young children on rough grass made entirely without a lathe – the piston being filed to fit the by no means perfectly round cylinder. So, there is no reason at all why those readers who have yet to acquire a lathe should not build successful engines; the first two described, anyway, for I would regard a machine as desirable for the crane and essential for the final, miniature, steam plant.

Chapter Two
POLLY
A VERTICAL STEAM PLANT

It is something like 60 years ago that I was given a little engine like this one, and I can still remember the excitement when the box was opened. It gave me pleasure for very many years (and my father too, I believe!) and despite the fact that I now have quite a collection of more sophisticated models, and have had the opportunity of handling "real" steam engines many times in the intervening years, I still get some satisfaction from running a little engine of this type. I think it must be the sheer simplicity of the "works", coupled with the engine's general air of busyness when it is running!

In designing this one I have tried to keep to the general lines of the original, Fig 2–1, but I have made one change of some importance. All engines of this type with oscillating cylinders have a single port in the cylinder which communicates in turn with the inlet and exhaust port in the steam chest. This one has two ports in both cylinder and chest, which gives two advantages. First, the accuracy of spacing needed with the single port design is no longer necessary; the ports are bound to line up properly. Secondly, it offers the opportunity of experiment in (e.g.) making the steam and exhaust ports of different sizes. I haven't done this on the drawing, but you can, if you wish, try making the exhaust port with a No. 50 drill instead of the No. 53 shown. This would give you almost 35% greater passage area. The only penalty you pay for this arrangement is that the portface is a little wider than usual for a cylinder of this size.

I have designed the boiler with a centre flue. Many of the engines of that era had just a simple pot (or tank) boiler with very limited heating surface and, moreover, the flames, often licked out from the firebox round the outside of the boiler shell. This not only destroyed the polish but could be a bit dangerous as well. If you wish to direct the exhaust pipe up the chimney you can do so, but though it adds a bit of draught, and provides the desirable drift of steam from the top, I think the appearance of the engine is spoiled a bit – the external pipe destroys the simple lines of the plant.

A little consideration of the drawings shows that you can't make the engine part till you have made the standard which carries the steam chest; you can't finish the standard till you have the firebox; and you need the boiler shell finished before you can start on the firebox. Therefore – start by making the boiler shell. (Common sense, I know, but Production Engineers actually use Computers to solve this sort of argument!)

Boiler Shell Plate Fig 2–2.

If you have a piece of brass or copper tube of the right size, by all means use it, and

Polly

save the jointing operation. If you have a piece which is *almost* the right size, you can use that – but remember what I said about such alterations in the introductory chapter. Otherwise you need a sheet of 24 SWG (or 0.6mm) copper at least $6\frac{3}{4}''$ long and a little over $3\frac{1}{4}''$ wide. You can use brass for this little chap if you like; it will polish better, but isn't quite as easy to manipulate. File one long edge till it is dead straight, and remove burrs. Mark out to $3\frac{1}{4}''$ wide and cut and file to this line. It is more important that the two edges be parallel than that they be exact to dimension. From the first edge you filed mark out a line with your square close to and across one end. Cut or file to this line, checking with your square as you do so. Remove burrs.

Now lay the sheet on a flat surface and

A Vertical Steam Plant
Fig 2-1

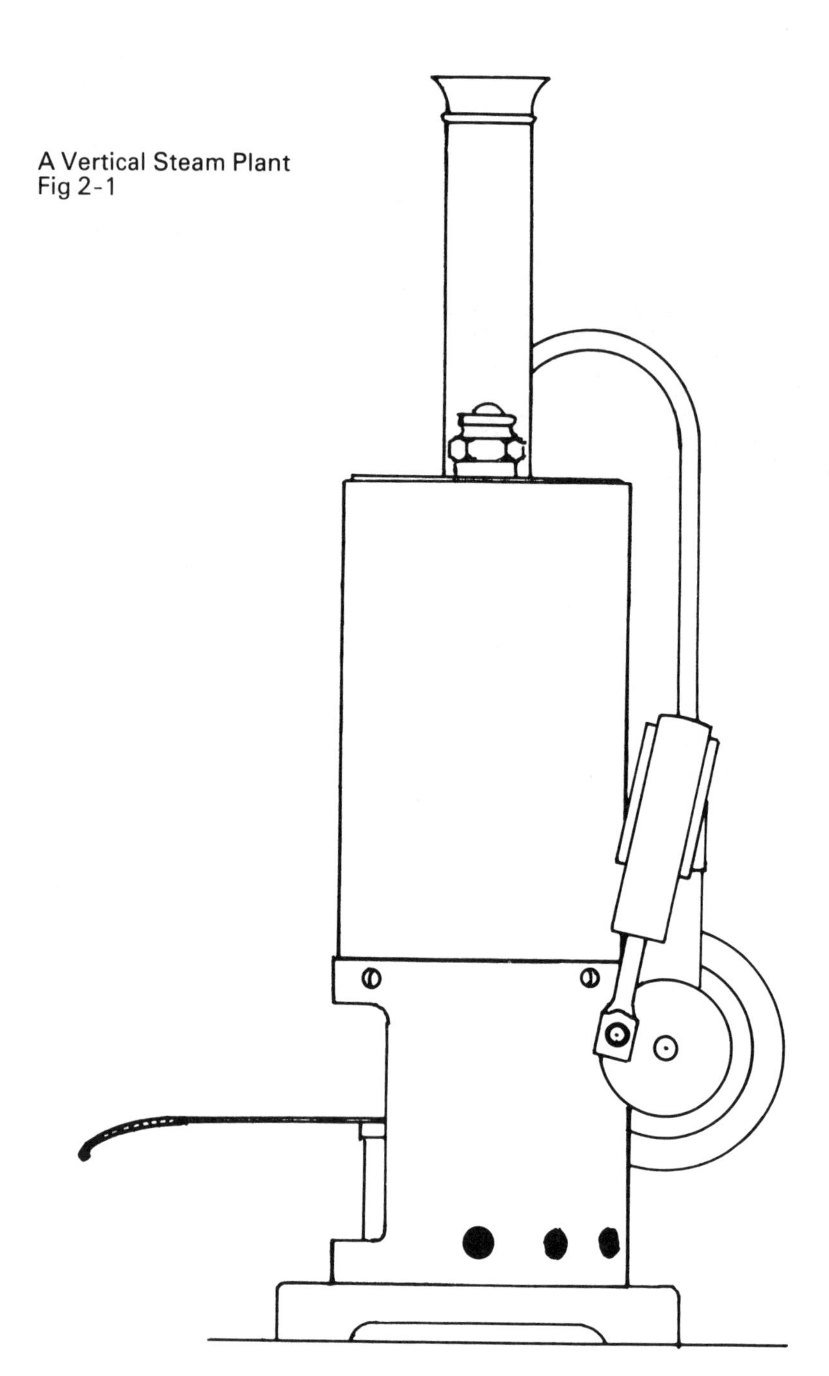

Boiler
Fig 2-2

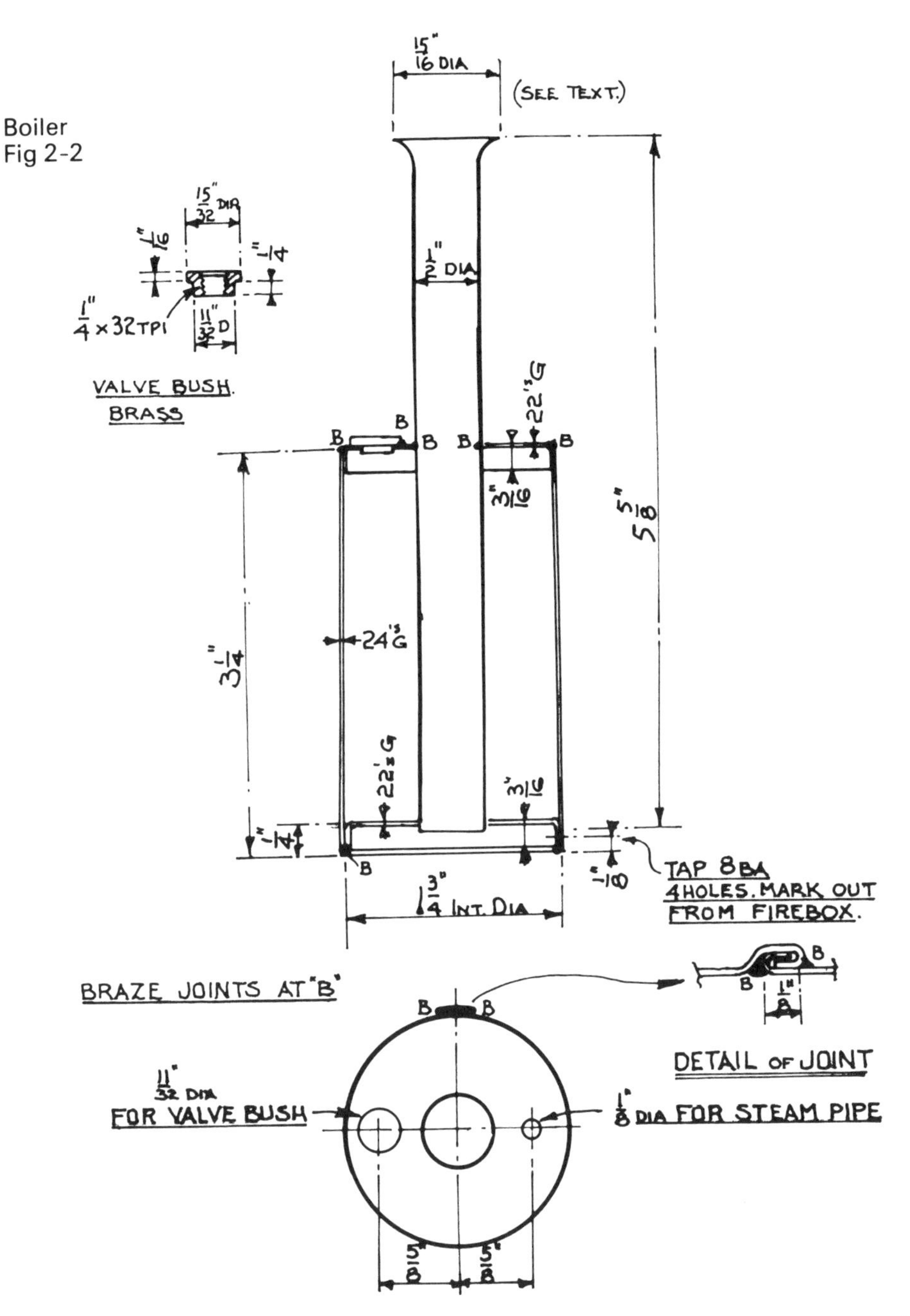

butt this squared-off end against a fairly substantial chunk of metal. Set your steel rule against this same chunk and mark out a distance of 6–1/32″ – this is the circumference of the shell with an added allowance for the joint. Square off and cut off as before. Check that both ends are square to the side you originally filed straight – this is what we call a "reference face", from which all other work is measured. (Those accustomed to woodwork will recognise it as a "face edge") Remove all burrs and sharp corners – just take the sharpness off, don't make a bevel. You will need to anneal (soften) it, but leave this job till you have prepared the endplates so that all can be done at once.

End Plate Material

These are thicker than the shell, for strength reasons. You need 22 gauge (0.7mm) about $2\frac{1}{4}$″ x $4\frac{1}{2}$″. But if you are using tube for the shell, larger than shown on the drawing, allow for two circles $\frac{1}{2}$″ larger in diameter than the bore tube. Mark out two circles 2–7/32″ dia, with a small centrepop at the centre. Mark out a centreline across one of them. Scribe two lines at $\frac{5}{8}$″ radius on this diameter and centrepop the intersections. Drill a 9/32″ hole in the centre of each. DON'T drill thin sheet hand-held in a drilling machine; if you can't clamp it to the table, use a hand drill in the vice. Flatten the sheet round the holes if they have turned up and cut round the circle, filing to the line. Remove burrs. Don't drill the other holes yet.

Now for the annealing. Fill the blow-lamp and make sure you have a pricker handy. Light up and get a good clean flame. Rest the three pieces of copper on some soft asbestos, or on "Fossalsil" *** insulating firebrick. Don't use asbestos cement – this won't stand flame without exploding! Heat the plates to red heat.

***To be had from Builders Supply Merchants.

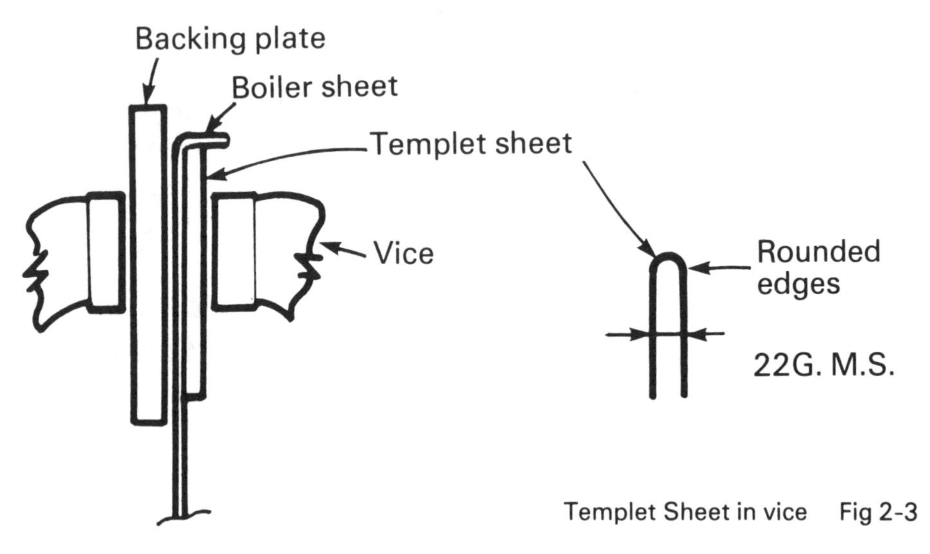

Templet Sheet in vice Fig 2-3

How red is red? In this case, a visible red, but not as red as a pillar-box.Now, it is *not* necessary to quench in water to anneal copper, despite what you may have heard. But quenching saves time and, in addition, helps to remove the scale formed on the surface. Note, don't handle the hot sheets too roughly, and if you do quench, lower the sheets into the water on edge, not flat.

Forming The Barrel

Mark out at each end of the rectangular sheet a line square across and 5/32" from the end, but *on opposite sides* of the sheet at opposite ends. Either fit smooth jaws to the vice, or use two pieces of clean and sharp-edged angle between the normal jaws and grip one end of the sheet with the scribed line aligned with the edge of the jaw, the line just visible. Use your square to make sure the sheet is upright. Fold the sheet to a right angle, the scribed mark on the inside, and use a mallet or soft-faced hammer lightly to beat down to a sharp corner, but take care not to hit too hard or you will distort the sheet. Repeat for the other end, the fold being, of course, to the opposite side of the sheet.

You must now form these two angles into hooks, as shown in Fig. 2–2. Set the sheet in the vice as shown in Fig. 2–3. The "template sheet" is preferably of steel, the edge dead square across, slightly rounded, and, say, one gauge thicker than the boiler shell-plate. Very gently tap the bent-over end of the sheet till you have folded it over the template, and then flatten it down properly. Make a good job of this, and take care not to get dents and burrs – hammer marks will show. You MAY find you have to re-anneal the plate, but I didn't find it necessary. But better to anneal again unnecessarily than to work on hard copper – you need full strength at the joint.

Find a piece of "stuff" about $\frac{1}{2}$" less in diameter than the diameter of the shell. Choose the best-looking of the folds and call this the outside. Fold round this mandrel and hook the two ends together. Don't worry if the "tube" seems a bit flat – you are only making sure the hooks fit. If the two ends won't go together, open them out a bit. When satisfied that the joint will go right home take all apart. Clean the slots using new, not oily, emery and apply a thin even coat of "Easyflo" flux inside and out. Reassemble the joint and make sure the hooks are fully home. Slip over the mandrel and if you can, hold this by both ends across the vice. If not, you will have to grip by one end, and you must hold it really tight. Now; take care. You have to "dolly" the joint as shown in Fig. 2–4, and you may find one end comes undone whilst working on the other. This must be avoided; if you like you can put in a small rivet (say 1/16") at each end first. With light blows from a soft-faced hammer or mallet close the joint end to end and then dolly down as shown in the sketch; the effect of this is to give a flush face on the inside. The dolly itself is just a piece of wood with slightly rounded edges. You will squeeze out a lot of the flux – not to worry, because the job will also get dirty! Clean down with a damp cloth, inside and out, and then apply a fillet of flux to the joint –

"Dolly" the joint Fig 2-4

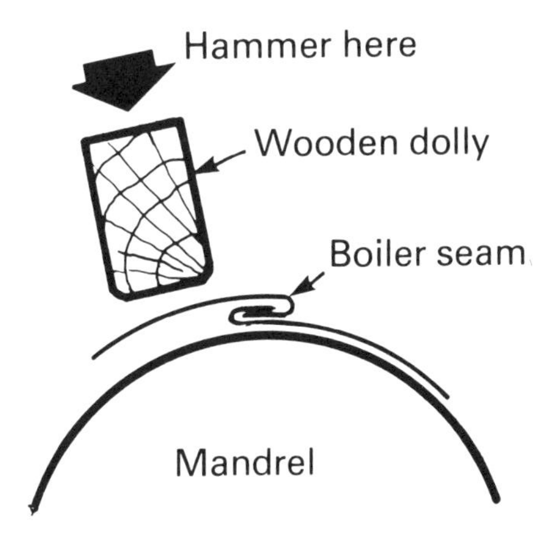

again both inside and out. (If you have used rivets, file off the projection)

Brazing The Barrel

If you have done no silver soldering before, you will need some preparation. A little job like this can be done in a disused fireplace, but I don't advise that in the spare bedroom – or anywhere else with a carpet! (James Nasmyth, inventor of the Steam Hammer, made the castings for his first engine thus, and got into trouble!) I use three 9" x 9" firebricks set in the shape of a letter "U" with a 9" x 12" across the back – these are quite cheap from builder's merchants. You need a bit of asbestos millboard on the bottom brick to stop it from scratching the work, and I suggest another sheet below the hearth if it is set on a wood bench; take care there is nothing behind the hearth to catch fire, too. A few odd bits of broken brick – or, better still, the Fossalsil I mentioned earlier – come in handy to act as props. As to heat, you need a $\frac{3}{4}$ pint or 1-pint paraffin blowlamp, or a propane torch of equivalent size. The small gas blowlamps – Soudogaz and similar – are not quite big enough for this job, though I do use them a lot for smaller work.

You will need about 18 inches of $\frac{1}{16}$ in. or $1\frac{1}{2}$ mm Easyflo silver solder – have two in case you drop one! A small pair of tongs; test-tube tongs from the chemists are fine (also very handy for turning bacon in the frying pan; get two, to keep the other half happy). Easyflo flux powder, and a little flux paste. Mix some of the powder to about the consistency of mustard with clean water. An old fishpaste pot is best to mix and keep it in, but make it fresh every time. Clean the blowlamp and fill it (yes; *again.* Always start with a full lamp) and have the pricker ready. Remove all wood shavings, aerosol containers, paint tins etc. from the vicinity, and have a bucket of water just in case. As to acid prickle, for removing flux after brazing, this is not essential, but it does help. Although I have been brazing for nearly 30 years I still get my acid ready diluted from the chemists. Concentrated sulphuric acid is nasty stuff, and the less I have to do with it the better.

A "Wincester Quart", (about 2 litres) costs only a few "p" – less than the bottle. The strength required is $2\frac{1}{2}$%, or one part of acid to 40 of water. You will need an acid-proof container, and for little jobs I use an old car battery with the inside divisions removed, but an old glazed earthenware water jug with no handle served me for many years. If you *are* going to pickle, add a pair of goggles to the gear. Those you use when grinding tools will do. Do the pickling well away from machines, preferably outside the shop.

This all sounds very formidable, but after a while you will find that silver brazing, provided you keep everything clean, is easier than soft soldering, especially if you have a propane blowlamp that does not need warming up.

Barrel joint

Smear paste flux all along the grooves both inside and outside of the shell. Prop it up in the hearth at about 30 deg. to the horizontal, joint at the bottom, the "up" end towards you. Make sure it can't roll. Light the blowlamp and pump up well. Warm one end of the silver solder and dip in the powder flux, coating it about 2 in. long. (Repeat this at intervals during brazing – never use solder with no flux on it). Now, gently warm the job to dry out the flux; do this indirectly and don't be in a hurry. As soon as it has dried, apply stronger heat from below. Keep the flame moving if you have used brass sheet especially, for you can easily melt brass with a blowlamp. As soon as the metal approaches dull red bring the heat to the nearer end of the joint and keep trying the end of the solder rod on the inside. As soon as it melts and runs into the joint, move the heat steadily down the joint outside and follow with the rod inside. You should see a nice bright line run down following the heat.

Note that it should not be necessary to exceed dull red – don't do the job in a bright light – excessive heat will "boil" the solder and give pinholes. Now turn the job over. Rub the fluxy rod down the joint on the outside to add a little flux. Apply heat as before, but from the side this time, and as soon as the joint is hot enough try the rod, diverting the flame momentarily; if you try to pass the rod through the flame you will just melt the end off! Heat down the joint as before, following with the rod. Finally, play the flame back and forth once or twice to distribute the solder evenly. Allow to cool to black heat, put on your goggles, and standing well back dump the shell on its end into the pickle, if used. If not, simply quench in clean water. Leave it there for half an hour to allow time for the scale to be dissolved. Remove, wash in running water, wirebrush with a brass brush or a pan-scrubber, and examine the joint for burning or pin-hole effects. If these are bad, file out the offending part, reflux the whole joint, and resolder as required.

You will probably find, if this is your first attempt, that there are patches of silver-solder all over the place. This is due to fluxy fingermarks on the material, and to unsteadiness in holding the rod. It can also happen if a patch adjacent to the joint gets hotter than the joint – the solder tends to follow the heat. Don't worry—you will improve with practice, and you can remove the patches with a fine file and polish the marks away when the boiler is finished.

End Plates

Whilst the shell is pickling, look for some material to make the former etc. Fig 2–5. The former and backing plate are best made of steel, the latter about $\frac{3}{16}$ in. thick. The length of the former should be chosen to suit the length of the bolts to hand, but not less than $1\frac{1}{2}$ in.

If you have no steel available, really hard wood will do for the former, but make it rather longer. An assembly of large washers will do for the backing plate.

Return to the shell. Get it as round as you can with your fingers, and then very gently bring it truly round by threading it on the mandrel you used before & gently tapping all over with the mallet, rotating it on the mandrel as you do so. Light blows should be used. Now measure the internal diameter of the ends. Chuck the former in your 4-jaw and set true – you can use a 3-jaw if the material is round stock. Face the end, aiming at a fine finish. Machine the diameter D to the dimension shown in the sketch. Fig. 2–5 Radius the corner. With the tailstock drill chuck, centre deeply and drill $\frac{9}{32}$ in. Use plenty of cutting oil and don't force it. Lightly countersink the hole. Polish all with emery. File a flat each side of the former. Assemble the parts as shown, and grip in the vice by the flats.

Now work round the end-plate with a small rawhide or plastic mallet, the blows almost vertical to start with, rapid blows, never in the same place twice, round and round. As the material bends over, increase the angle of attack of the mallet. When the material stands at an angle of about 45 deg. it may need softening again. Do this, and return to the beating, never with heavy blows – you must "persuade" the metal, not force it – until the material has followed the shape of the former. Finish off by going round the cylindrical part several times with many, fast, very light blows. Remove from the former. If the other end of the shell is within $\frac{1}{64}$ in. of the same diameter, you can repeat the operation, but if smaller, return the former to the lathe and skim down, and then beat out the other endplate. The flange may be a bit uneven; if so, trim with a file. During the whole of this operation, make sure that the backplate and former, as well as the endplate, are quite clean, as bits of dirt will make ineradicable marks on the soft metal.

The next operation is to open up the centre holes to fit the chimney. Make sure that the ½ in. tube is round and free from kinks, and cut off a piece 5¾ in. long. Trim the ends and remove burrs. Polish with very fine emery, but not Brasso at this stage. Measure the O.D. of the tube. Mount the first endplate in the three-jaw, gripping it inside the flange with only moderate force. Enlarge the hole using drill after drill rising $\frac{1}{64}$ in. at a time and then finish off with a boring bar so that the tube is a good fit, but not so tight that it won't slide. Clean up and polish with fine emery. Repeat for the other endplate. If you have no lathe, enlarge with a round file and finish off with a drill.

Now drill the holes in the top plate for the steam pipe and safety valve. You will have to use a commercial safety valve if you have no lathe – obtainable from Stuart Turner Ltd, Henley-on-Thames, Oxon RG9 2AD, or from A.J. Reeves & Co, Holly lane, Marston Green, Birmingham, and you should ask for one to blow off at 20 lb. sq. in. The Stuart valve used on their "S.T." marine steam plant is just right, and you should ask for it to be supplied complete with bush. This must be measured before drilling the hole in the boiler to accept it. Fig. 2–15 shows the "home-made" valve.

Mount the plate on the former in the drilling machine, use plenty of oil, about 2,500 r.p.m. for the small hole and 500 for the large one. As the drill point will come up against the former as you go through, there should be little risk of snatching. Remove burrs. Now make the safety-valve bush. Chuck a piece of ½ in. material in the three-jaw, face the end, and turn down to $\frac{15}{32}$ in. with a good finish about $\frac{5}{16}$ in. long. Centre, and drill $\frac{7}{32}$ in. for ½ in. deep or so. Grip a taper ¼ in. x 32 t. tap in the tailstock chuck, slide the tailstock up to the

End Plates Fig 2-5

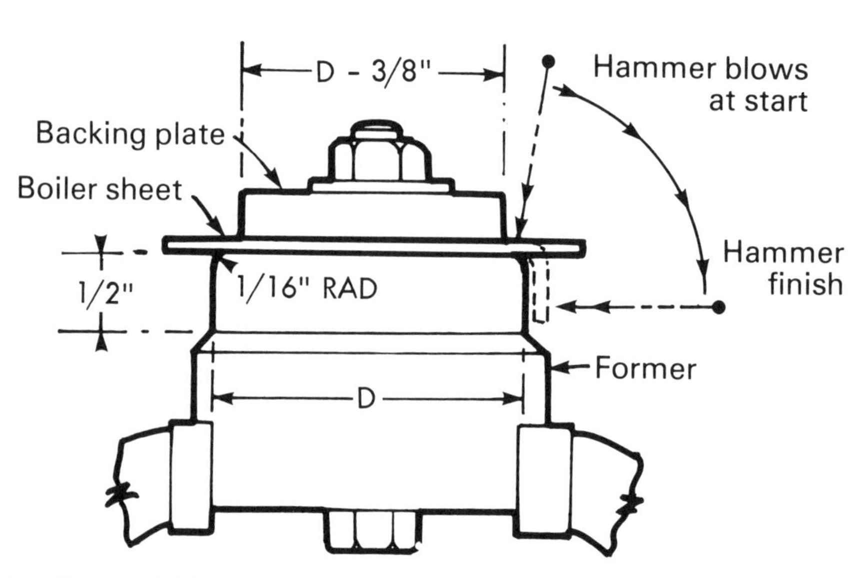

D = Bore of largest end of the tube LESS twice the thickness of the sheet.

work, and pull the work round by hand, rocking back and forth, whilst exerting a steady pressure by pushing behind the tailstock – don't use the feedscrew. When the tap has bottomed, follow with the plug tap. Very lightly countersink the hole. Now, with a knife tool, very sharp and running at top speed, turn down the $\frac{11}{32}$ in. dia. until it is a tight fit in the hole in the boiler endplate. Part off; reverse in the chuck with a light grip, face the end lightly if need be, and countersink the hole $\frac{1}{32}$ in. Run the plug tap through again to clean up the threads. The bush must be a tight fit in the plate and you can wit!. advantage make the O.D. a slight taper.

Chimney

Forming the top of the chimney is very much a matter of taste. "Real" boilers like this had the flue terminated with a semicircular band; you can do this with $\frac{1}{16}$ in. half-round brass strip soldered on. You can serrate the top, like Mr. Stephenson's Rocket; you can turn a nice brass end-piece to slip on. That on the prototype was beaten to shape. Practise on a spare piece of tube. Turn a bit of a radius on the end of 5 in. of $\frac{7}{16}$ in. steel rod, and grip firmly in the vice. Anneal the very end of the tube, as for the boiler plates. Hold in the hand over the rounded end of the mandrel, at a very slight angle. With the ball end of a small riveting hammer, tap with fast, light blows on the very end, rotating the tube as you do so. Note, you are NOT trying to bend the tube to shape, but to thin the material. As it gets thinner, it must get larger in diameter; it can do nothing else. So, gradually, the curve will form. As it does so, move the point of attack slightly back along the tube. Then increase the slope of the tube in relation to the mandrel, and return to the edge. Work on like this until you have the shape to suit your taste. Re-anneal from time to time. To finish off, do some selective beating here and there to get the end symmetrical, always holding the curve tangential to the mandrel where you are hitting it, and finally, chuck in the three-jaw, pull round by hand whilst you trim the end with a knife-tool, and finish by polishing out all hammer marks.

Assembly

Now to assemble the boiler. The endplates may need a little work with a file to make them fit, but they should be tight. Make sure all parts are clean and free from oil. Have three pieces of silver solder ready, as before – don't use very long bits, as they are difficult to control; 9 to 12 in. is enough. Smear flux on the safety valve bush and press it in, being careful not to bend the plate. Flux the flange of the top plate and the inside of the barrel and fit this. Take some trouble to see that the holes are in the right position in relation to the seam – see drawing. If there is any suspicion that the plate is going to shift when it gets hot, drill 3 little holes – say $\frac{1}{16}$ in. – at 120 deg. intervals through flange and barrel, poke in $\frac{1}{16}$ in. rivets and just give them a tap with the hammer. The ends can be filed off after brazing. Do the same with the bottom plate, putting a trace of flux down the seam on the inside first. The edge of the flange should be about $\frac{1}{16}$ in. inside the barrel. Put a smear of flux inside the chimney holes, and slide the chimney in from the top. Carefully arrange narrow rings of flux: round the chimney, top and bottom, round the joint between plates and barrel, on each rivet, on the bush, and along the seam. Wipe off all surplus elsewhere.

Set the boiler upright, with a little piece of asbestos under the lower end of the chimney to stop it from sliding down under heat. Lean two or three pieces of asbestos round the shell, to reduce heat losses. See that everything is to hand, as before and warm the job to dry the flux, fluxing the rod at the same time. Now raise the temperature around the chimney, keeping the flame on the move, until the rod runs

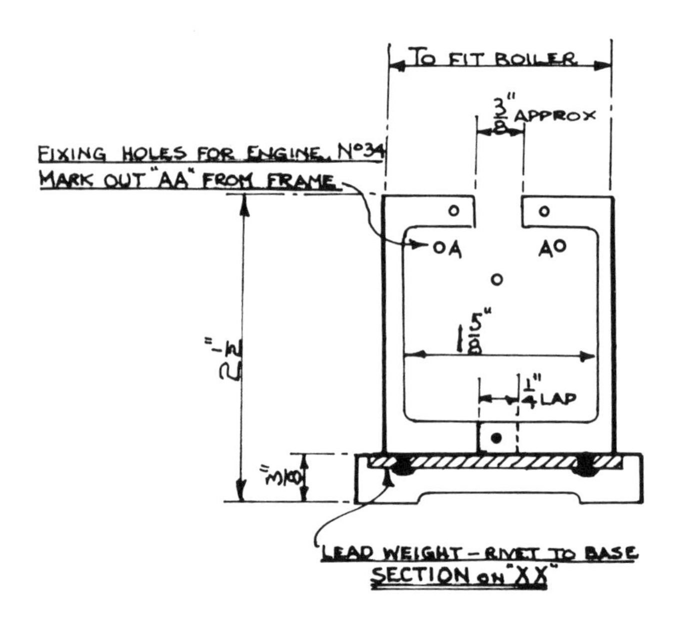

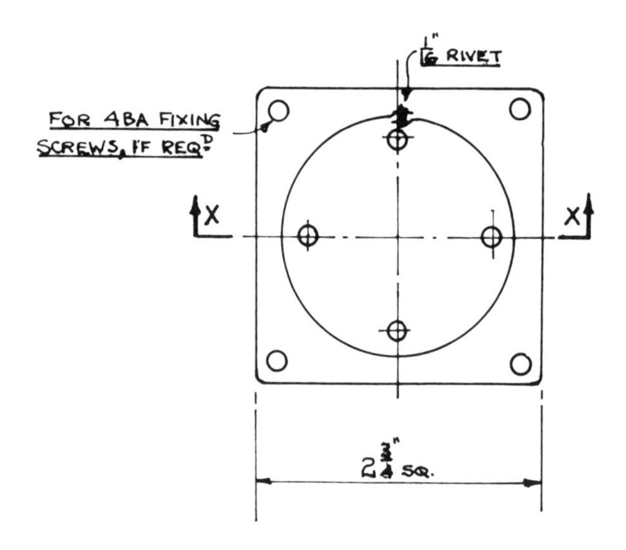

Firebox and Base Fig 2-6

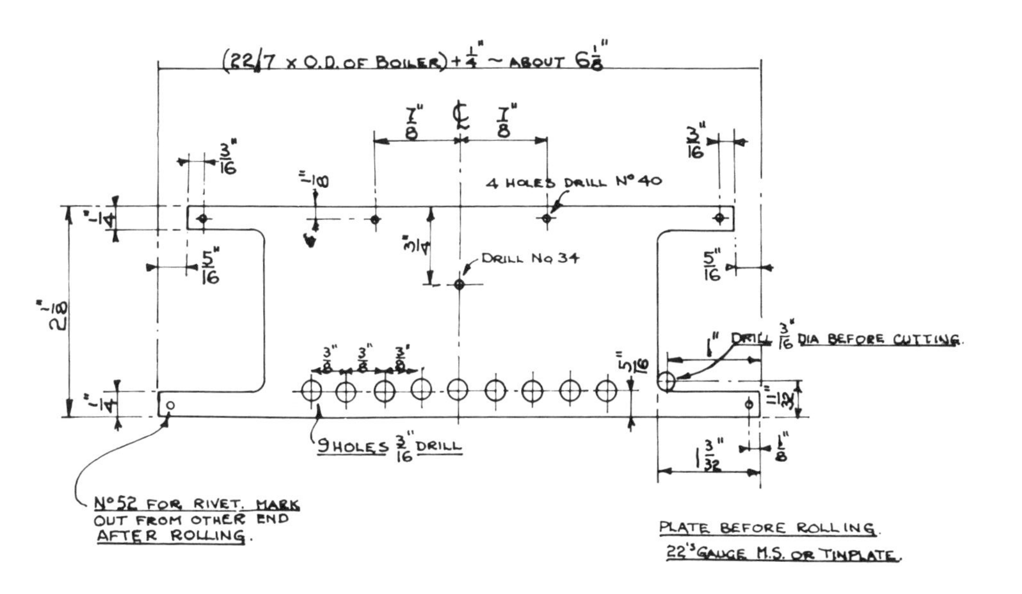

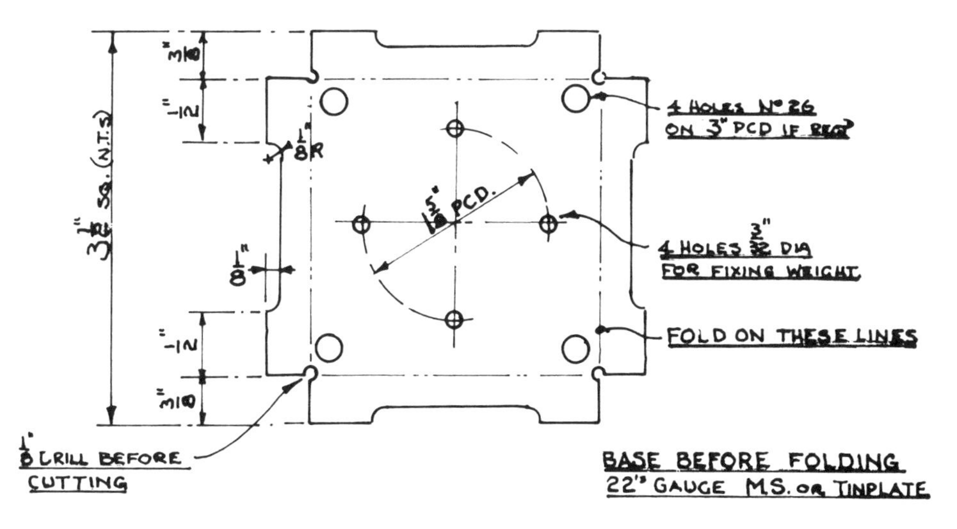

Firebox and Base Fig 2-6

as before; take the lamp a circuit round, adding more rod if need be, till you get the bright ring you will now recognise. Move the flame to the bush and braze that–don't get it *too* hot or it will melt; dull red is enough. Then braze up the outer joint. This will need quite a bit of rod, so have the second piece ready fluxed. Again, let the flame go round the full circle, adding rod as needed. Touch each rivet with the rod as you pass.

Allow to cool a minute or so, till it is safe to handle with tongs – hot brass or copper is very fragile – and turn the job over. If you find that asbestos has stuck to the flux – and it usually does! – let it cool further and wash it off. Then re-flux all joints. Prop the job so that it is secure, and not standing on the chimney, apply the heat shield again, and repeat the brazing operation on the bottom. Cool to black, and dump in the pickle, chimney side up. Shake it about under the acid to make sure that it gets inside. Leave for 20/30 minutes. Wash thoroughly under the hot tap.

Remove any stubborn pieces of flux you may see, and scrub with a brass wire brush or pan scrubber.

Now thread a short length of pipe to fit the safety-valve bush, making the thread about $\frac{5}{16}$ in. long. Screw this in, and attach a piece of rubber tube. Hold the boiler under water, with your finger over the small hole, and apply pressure. (Blow into it if you haven't a car tyre pump!) Turn the job about, and look for leaks. Mark any you see with a felt pen. Any serious leaks, or if there are a lot, means you must reflux all joints and rebraze the offending parts. But the odd pinhole in (say) the bottom plate can be caulked with soft solder. Dry the boiler, file off the projecting rivets and any bits of stray brazing material, and then clean up the file marks with fine emery. Rub over with fine steel wool, and then polish. She's a boiler!

Readers will think that this must take a long time! In fact, it has taken four times as long to describe as it took to make, so don't get disheartened!

Firebox and Base Fig. 2–6

If uncoated steel is being used, remove all scale or rust with wire brush and emery. File one edge of the sheet truly straight and remove burrs. Check the boiler diameter at the bottom and work out the length required as shown on the drawing. With square, scriber, and dividers, mark out and cut to length and width. Mark out and centre-pop the position of all holes. With dividers scribe the four $\frac{3}{16}$ in. circles at the corners of the cut-outs. As the plate is fairly large it is safe to hold down by hand on the drilling machine; rest it on a piece of hard wood and drill, using a light rate of feed, and cutting oil. Remove the burrs from the holes. Mark out the cut-out, tangential to the holes in the corners. Working inside the lines, make straight cuts, into the holes, with the tin-snips, and follow with diagonal cuts to remove the bulk of the material. With a little "wangling" you should be able to get most of it out with the snips, but don't bend the sheet about too much. Finish to the lines with a fine file; remove the sharp corners and burrs.

Roll the sheet round a piece of wood or bar of about the right size, making sure that all is square. Wrap a piece of soft wire round to hold it, with a twist at the end, and offer up to the boiler. Tighten or loosen the twist till it fits sweetly. Mark through the No. 52 hole at the bottom onto the other limb; remove the wire; centrepop the mark, and drill the hole. Pop a $\frac{1}{16}$ in. rivet through from the inside, rest the head on a bar in the vice, and neatly rivet over. Manipulate the "tube" with the fingers till it both looks right, and fits the boiler shell. Using either solder paint or Baker's fluid, tin the inside for about $\frac{1}{4}$ in. from the bottom. You will need at least a 60 watt electric soldering iron for this job if it is not tinplate, and I always use a large plain copper "iron" heated in the blow-lamp for such work. By the way, if you do this job in the brazing area, take GREAT care that neither flux nor solder drops get onto the asbestos you use for brazing. Either will be fatal to any later brazing jobs.

For the base, square off a piece of the same material to the size shown. Mark out, centre-pop, and drill all the holes. Mark out the corner cutout with a square, and those in the sides with dividers. Scribe a circle from the centre about $\frac{1}{8}$ in. larger than the firebox diameter. Cut out the corners etc. and clean up with a fine file. Remove all sharp edges.

Using the same technique as you did to fold the seam on the boiler, fold down at the dotted lines, taking great care that you fold actually *on* the lines, and that the job is square. Tin the top of the sheet to about $\frac{1}{16}$ in. inside the circle, and the inside of the corners. Solder up the latter, making a nice fillet inside. With a piece of wood across the top of the firebox, clamp the two parts together with a G-cramp – not too tight – making sure that the firehole door is aligned to the centre of one side of the base. With a good hot iron, solder the two parts together, doing all the work from the inside. Wash well in hot water, and dry off.

Offer the boiler up to the firebox, with the seam aligned to the firehole, and mark through two of the No. 40 holes diagonally opposite each other, the base of the boiler shell being about $\frac{1}{8}$ in. below the holes. Remove, drill the holes No. 50 and tap 8 BA. Be careful – use cutting oil and go very steady; 8 BA taps are fragile, and copper especially is sticky stuff to tap. Refit, with short screws in these two holes, and square up the boiler to the base. Have a good look at it, as well as using the square – you don't want a boiler like Chesterfield church spire! Mark out and drill the other two holes in the shell, tap as before, and fit screws. Have another good look at it – show it to the family (AREN'T you clever, dear! Can you fix my sewing machine?) and if it is at all out of square, elongate ("draw" is the trade term) one of the holes in the firebox to put it right.

Take off the firebox and clean off all fingermarks with "Thawpit" or cellulose thinners, and give it a couple of coats of cellulose primer – preferably the red sort in aerosol cans. Allow to dry well, and then a couple of coats of matt black. Cellulose is best, as it seems to stand the heat well, and the aerosol cans make the job very easy. Set it aside and don't touch for 24 hours. The top of a radiator is a good place to harden the cellulose, but make sure it has "set" first.

Engine Standard Fig. 2–7

Cut a piece of material $2\frac{1}{8}$ in. wide and $3\frac{1}{8}$ in. long and square up two adjacent sides with a file. Take care over this, as these sides are your reference edges; mark them with a felt pen. Set out the two centre-lines and make a pop mark where they cross. Set your dividers to $1\frac{9}{16}$ in., mark for the No. 33 hole and pop that. From this pop mark, again using dividers, mark and pop the three No. 52 holes. Set dividers to $1\frac{1}{4}$ in. and mark the vertical centre-line on the lower limb – it has one No. 34 hole on it and lightly pop a mark where the two lines cross. From this pop, mark out the other $\frac{1}{4}$ in. hole at $1\frac{1}{4}$ in. radius; the two folding-lines at $\frac{13}{16}$ in., setting these up from your reference edge with a square, and the three No. 34 holes. With dividers, scribe a short line $\frac{9}{16}$ in. from the left hand edge, to cross another one $\frac{5}{8}$ in. from the bottom edge, and pop. This is the centre of the circle that forms the radius in the "elbow" Now scribe the lines for the $\frac{1}{2}$ in. and $\frac{9}{16}$ in. widths of the limbs, and, finally, square off lines for the ends of the limbs.

Don't do any cutting yet, for we want to keep as much strength in the sheet for as long as possible. Deepen the pop marks for the two $\frac{1}{4}$ in. and the one No. 33 hole a little with a small centre-drill – don't do more than countersink them. Rest the

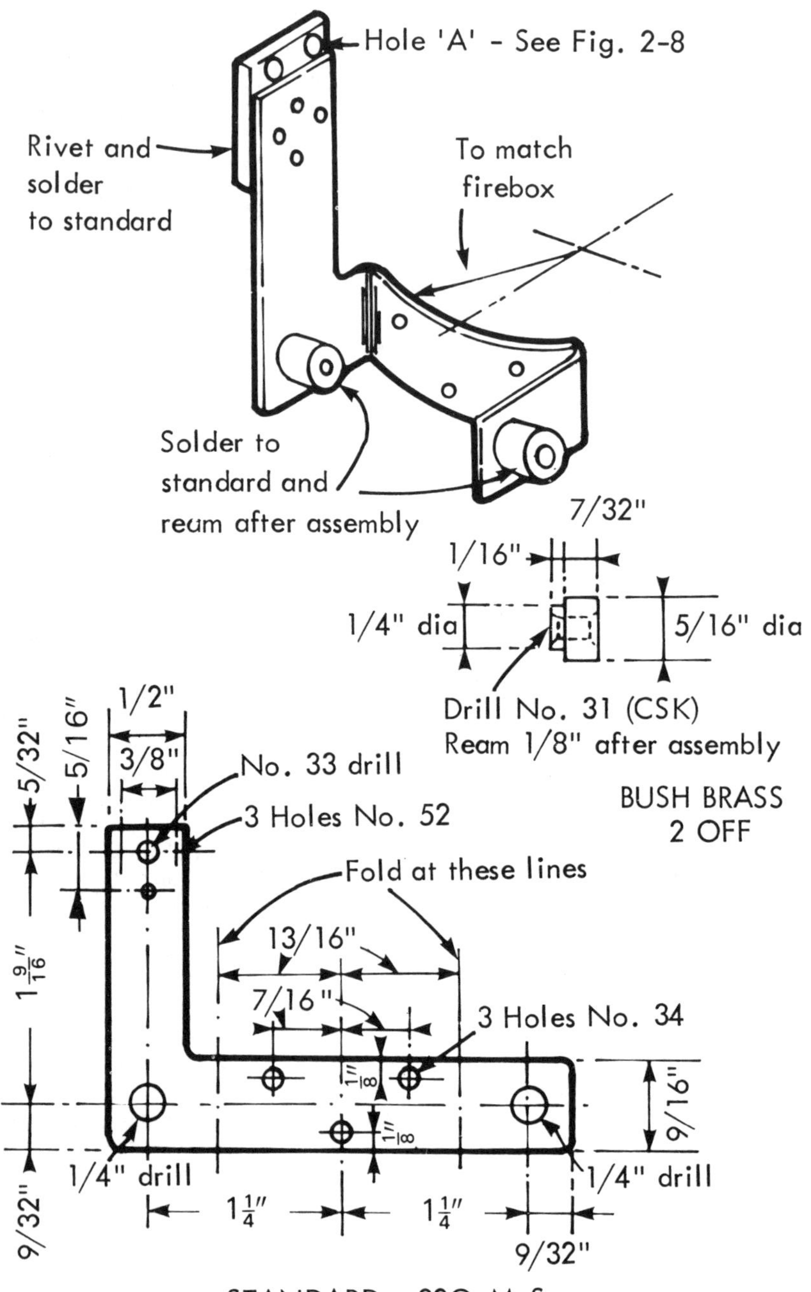

Fig 2-7 STANDARD - 22G. M. S.

sheet on an old piece of steel plate, carefully line up a $\frac{1}{4}$ in. drill to the left-hand hole, and clamp down. Using a slow speed and cutting oil with a steady feed, drill this hole. Similarly for that at the other end. (However, if you are building the engine and have no lathe you cannot use bushes. See the note at the end of this section) Change for No 33; again line up and clamp, and drill this one. The position of these three holes is important – a case where you start again if anything goes wrong, so take pains over them. The other holes may be drilled in the normal fashion, together with a No. 30 or $\frac{1}{8}$ in. hole in the "elbow". Remove all burrs on the holes. You can now cut along the previously marked lines to form the L shape, trim off the ends, and remove corners and sharp edges. If the sheet has twisted at all, use the mallet on a flat plate to straighten it.

With smooth jaws in the vice, set up dead square and fold the frame with as small a radius as you can manage, but not dead sharp. Use a small fibre hammer if need be, though a steel riveting hammer can be used if you are careful. Check for squareness, and make sure you have it the right way round! (The ends fold forwards towards you when the upright limb is on the left.)

Now form the radius on the back. You can do this on a piece of wood the same size as the firebox, and it is better to have it a shade small than a shade large. This will splay the ends out, and you must bring them back parallel to each other with patience and pliers. That will have altered the radius of the back, so repeat again! Finally, poke a piece of $\frac{1}{4}$ in. rod through the holes and twist until the rod is square to the upright and parallel across. Clean up with a wire brush, and tin around the $\frac{1}{4}$ in. holes, on the side the bushes project, and also the top $\frac{5}{8}$ in. of the upright, where the portface fits. See the drawing. Fig. 2–7.

If you have no lathe, then proceed as follows. Find a piece of brass tube which is an easy fit to your crankshaft – i.e. $\frac{1}{8}$ in. – and measure the outside diameter. Drill the holes in the standard this size instead of $\frac{1}{4}$ in. Cut off pieces of this tube about $\frac{5}{16}$ in. long and file the ends flat. Solder these in place in the standard with a good fillet of solder as you have no spigot or shoulder; to get them straight, fit a piece of $\frac{1}{8}$ in. rod through the hole whilst doing the soldering. You can now ignore the section which deals with the bushes!

Bearing Bushes Fig 2–7

You can now get back to the lathe, so chuck a piece of $\frac{5}{16}$ in. dia. brass, about $\frac{1}{2}$ in. projecting, and polish the outside. If it looks rough, turn down a piece of $\frac{3}{8}$ in. stuff to diameter. Face the end, and with a sharp knife tool turn down the little shoulder. Make it a good fit in the hole in the standard. Carefully centre the end with a small centre-drill; change for a No. 31 drill, and at your fastest speed and with very light feed on the tailstock screw, drill $\frac{5}{16}$ in. deep. Clear the drill frequently – the hole must not run out. Lightly countersink. Part off to make the bush a shade over dimension. If there is any sign of the drill point in the piece left in the chuck, face it away. Now repeat for the second bush, but fit it to the other hole in the standard. Finally, grip each bush in turn, lightly, in the chuck, and take a fine facing cut across. Countersink the hole a little. Set aside for the present.

Portface Fig. 2–8

You know a good deal about marking-out now, so the next, and most important, part should not worry you! For the portface you need a piece of hard brass. Normal sheet brass is a bit soft and you should try to get a piece of drawn brass flat if you can. This will almost certainly need filing on the faces, so start with a bit of $\frac{5}{32}$ in. File out the scratches and marks on both sides – no more; the thickness is not critical. Choose the better side, and make this

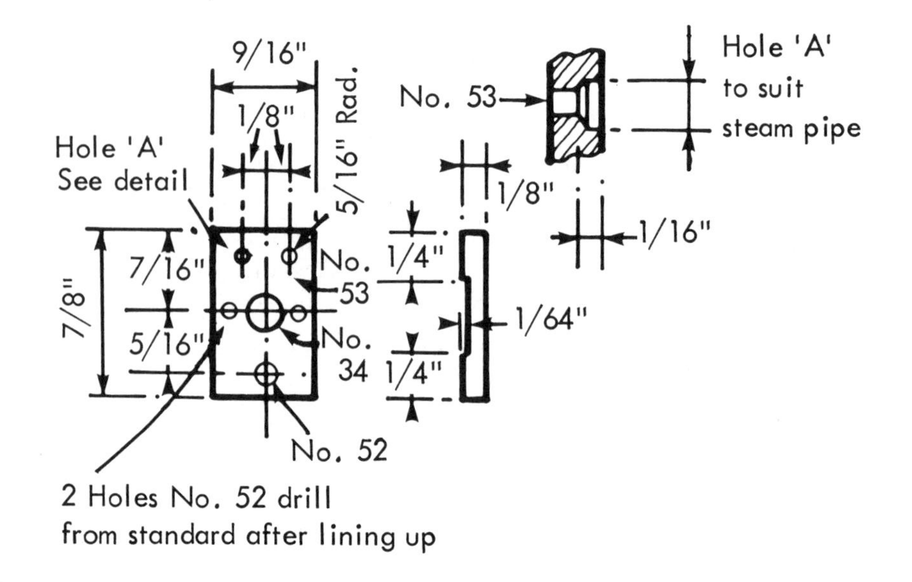

Fig 2–8 Portface – Hard Brass

really flat. Lay a piece of fine emery on a flat surface – the drilling machine table, say, (but NOT the lathe bed) – and rub around in circles with pressure from one finger in the middle of the piece. When you *think* it is flat, change to very fine emery, and repeat.

File one edge to form a reference edge. From this mark out the two centre lines, and pop the intersection. Now, with great care set your dividers to $\frac{5}{16}$ in. and draw a circle on a piece of spare brass. Measure the diameter. If it is $\frac{5}{8}$ in., well enough, but if not, repeat until the diameter *is* correct. Now scribe at this radius for the two small steam passages, and for the No. 52 hole at the other end. Set the dividers to draw a $\frac{1}{4}$ in. dia. circle, and from the intersection of the $\frac{5}{16}$ in. radius and the vertical centreline, mark out the two $\frac{1}{8}$ in. dimensions shown on the drawing. With a very sharp centre-punch, very lightly pop the holes, but you can go harder at the No. 52 hole at the bottom. Do NOT mark out the other two No. 52 holes. Carefully deepen the pops in the steam port centres by twirling a small drill in your fingers.

You can now, with square and dividers, mark out the outline of the block. But don't cut it out till you have drilled, thus saving lost work if anything should go wrong. (I had to make two of this part myself – poked the wrong size drill through the port!) Hold the block, with a piece of brass beneath, in a machine vice, and carefully line up a No. 53 drill to the first of the port holes. Run as fast as you can, and with steady and light feed, drill right through. Repeat for the other hole. Next, with a small centre-drill, go part way through the No. 34 hole, follow with No. 34 right through. Finally, drill the No. 52 at the bottom. Remove the burrs from the backside.

Turn the block over in the vice and, making sure you pick the right hole, set the depth stop of the drill so that a No. 31 (for $\frac{1}{8}$ in. steam pipe) will go just $\frac{1}{16}$ in. deep (see detail on drawing). If your machine has no depth stop, use great care and a hand drill. Remember, the drill will be greedy as there is a hole there already, so control the feed carefully.

You can now file to shape, but use soft jaws in the vice to hold it. Mount the block on the standard with a 5 BA bolt through the centre hole, and a $\frac{1}{16}$ in. rivet or a piece of wire through the No. 52 hole. Make sure it lines up square and tighten the bolt. With a hand drill, mark through the standard into the block for the two No. 52 holes not yet drilled. Return to the drilling machine and drill right through. Countersink all three holes on the front face of the block – the flat one is the "front". Finally, with dividers, scribe two lines across the front of the block $\frac{1}{4}$ in. from each end, and lightly file a relief between them. Remove all grease, and thinly tin the back. Wash off the flux (if you use solder paint for the job, scrub it in hot water) and clean out the solder from the holes – just poke the correct size drill through with your fingers.

Return to the very fine emery and flat surface, and rub the front of the block until the scriber-marks are removed. Coat the back with a thin and even layer of soft solder type flux, and ditto the face of the standard. Fit $\frac{1}{16}$ in. rivets in the two holes either side of the centre hole, roundhead on the standard side; snip off so that about $\frac{1}{32}$ in. projects on the portface side and give just one light tap with the ball end of a light hammer. Cut off and file the end of a third rivet so that it is just level with the portface when pushed through the third hole. Rest the standard on a flat piece of metal, and very carefully set the point of a sharp centre-punch in the centre of this third rivet; give one, sharp blow. Rivet the other two over in the usual way. File the two centre-rivets down flush with the surface. Heat the whole with a small blowlamp flame – this is where the little gas lamps come in very handy – until you see the solder melt. Allow to cool naturally, but do not quench. Clean off all flux residue.

Now, with a very fine Swiss file, or with fine emery wrapped round a small square file, bring down the third rivet level to the portface, if there is any projection. Clean any solder out of the holes, and from the edges of the block. Flux the bearing bushes and the standard and, one at a time, clamp them in place and sweat them. (See the note on page 00 if you are working without a lathe). Add a very little solder if needed to make a small fillet on the bosses. If your lamp won't give a very fine flame, you can clamp a piece of brass to the portface to act as a heat sink, to avoid melting the solder there. Clean off the flux.

Bolt the standard to the firebox with a 6 BA bolt and nut, and line up so that all is square; tighten the bolt, and carefully drill No. 34 through the other two holes into the firebox. Fit 6 BA bolts here. Push a No. 31 drill through both bearings and bend the standard until the drill spins freely – checking that the portface stays vertical. Ream right through both bushes with a $\frac{1}{8}$ in. reamer. If you haven't one, a $\frac{1}{8}$ in. twist drill will do. Remove the standard, and give the steel parts a coat of primer followed by colour to choice.

Crank Fig. 2–9

Back to the lathe! Chuck a piece of 1 in. dia. steel in the three-jaw, and machine down to a fine finish $\frac{7}{8}$ in. dia. and about $\frac{1}{2}$ in. long. About 400 r.p.m. with cutting oil. Face the end and then with a knife tool turn the boss, $\frac{5}{16}$ in. dia. and $\frac{3}{16}$ in. long. Bevel the corner. Aim at a fine finish from the tool – you should NEVER remove toolmarks with emery! Set a sharp pointed tool at dead centre height, facing the job. Bring it up to the boss, and wind the cross-slide a little past the centre. Bring back exactly to the centre of the boss, and note the index reading. Wind back a further 0.312 in., advance the point to the face

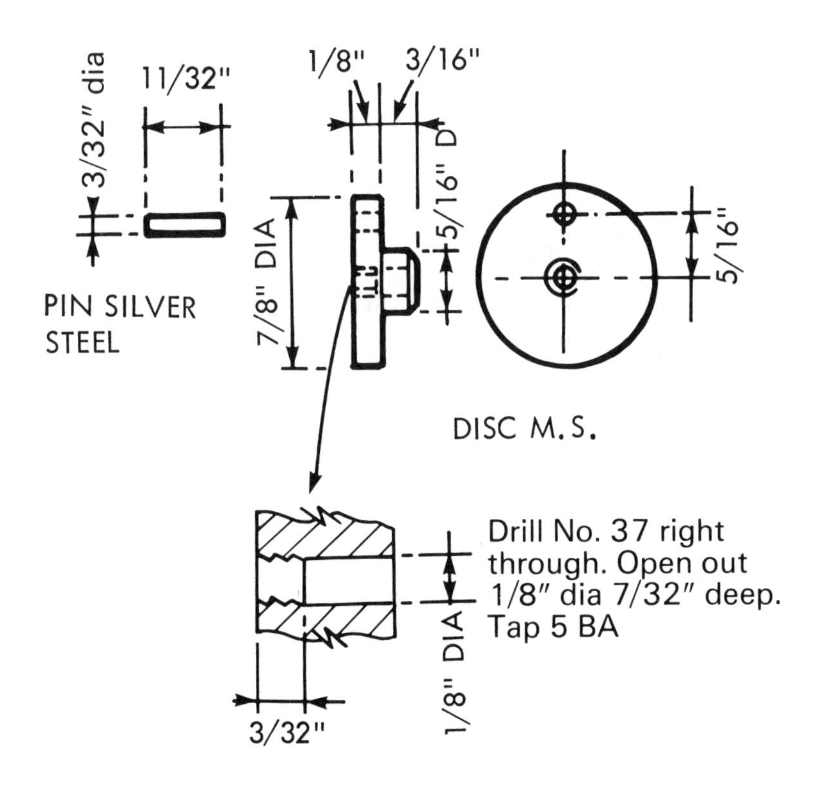

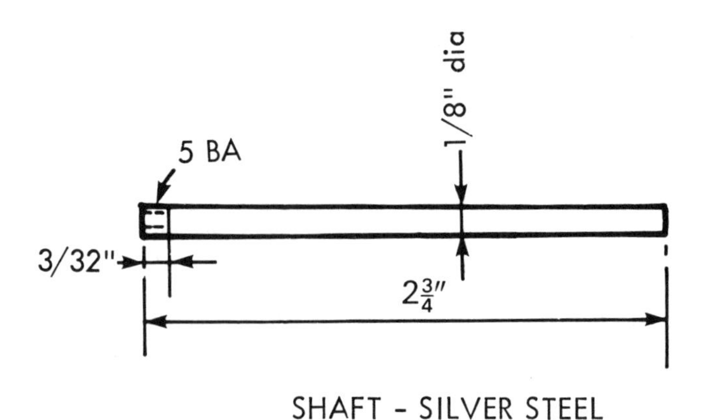

Crankshaft Details Fig 2-9

and pull the chuck over a few degrees to scribe a line about $\frac{3}{16}$ in. long. This will mark the centre for the crankpin. Centre the boss with a small Slocumbe drill. Running as fast as you can, and with light feed, drill No. 37 about $\frac{5}{8}$ in. deep. Change to $\frac{1}{8}$ in. drill, and open out $\frac{7}{32}$ in. deep. Carefully tap 5 BA; you can use your small tap wrench, as the $\frac{1}{8}$ in. pilot hole will keep it straight. Part off so that the flange will be a shade over the dimension. Hold in the chuck by the boss, lightly tap it back to the face of the jaws till true, and take a fine facing cut with a roundnose tool.

Remove from the lathe, and centre-pop on the line you scribed earlier, for the crankpin hole. Set up in a vice on the drilling machine, and drill No. 43, not forgetting to have a piece of metal underneath. Clean off the burrs, and run the 5 BA tap through the boss again.

I have dealt with crankshafts without a lathe in the first chapter. In this case, you ignore the $\frac{5}{16}$ in. diameter boss on the crankdisc, and use 5 BA washers to make up the distance. The end of the shaft should of course, be screwed for $\frac{1}{8}$ in., not $\frac{3}{32}$ in. as shown; if the end projects beyond the disc, file it flush.

To make the pin, chuck a piece of $\frac{3}{32}$ in. silver steel in the lathe or in your drill, and spinning at top speed bring down $\frac{1}{8}$ in. at the end until it just enters the No. 43 hole in the disc. Part off $\frac{11}{32}$ in. long (or use a saw if you must!) reverse in the chuck, and clean up the end with a knife tool. Remove burrs. Enter the little pin into the hole, and taking great care that it is square, gently squeeze it into the hole using the smooth jaws of the vice.

Shaft

Cut off a length of $\frac{1}{8}$ in. silver steel, and face one end in the lathe. Reverse in the chuck, and using a tailstock die-holder, cut $4\frac{1}{2}$ threads on the end. (If you have no such die-holder, make one! You can't do without it, and there have been some good designs shown in *Model Engineer*). Now grip the rod in the tailstock drill chuck, with as little projecting as possible. Set up the crank-disc in the three-jaw, boss outwards, and grip lightly. Spin the chuck, and tap the disc until you get it running true. Tighten the chuck, but not too much, as it does them no good to hold work at the extreme ends of the jaws. Advance the tailstock along the bed until the screwed end enters the hole, rotate the chuck slowly by hand (put a drop of oil on the thread) until you find the rod wants to rotate in the drill chuck. Remove from the machine, and you have a crankshaft.

Flywheel Fig. 2–10

The problem here may be to find a piece of stuff to make it from. (See the first chapter if you have no lathe). You need a piece about $1\frac{3}{8}$ in. long to allow for chucking; neither the width nor the diameter are important, but don't go above 2 in. or it will foul the boiler. Chuck the piece, using the outside jaws – 4-jaw if the piece is rough, in which case you need a piece of chalk and some patience to get it running true. Get a good hold of it. Face the end, and if you *ARE* using the 4-jaw, take a roughing cut until the projecting part is truly cylindrical. Change to the 3-jaw, and grip by the trued part; face the other end. Bring the outside to a fine finish (220 r.p.m.: for steel, with oil, and up to 300 r.p.m. for brass) with a slightly round-nose tool and fine cuts and feed. With the tool about 45 deg. to the work, remove the surplus metal to form the boss, until it is about 10 thou oversize and $\frac{1}{16}$ in. too long. Do this with the saddle locked, using the cross-slide.

Machining the recess will depend on the sort of tool you have available. I use a slightly tapered round-nose tool, set at a slight angle to the line of the lathe bed so that it clears the boss. (I have a 4-tool turret, and get the right position simply by adjusting this). Machine out the recess as deep as is needed and then, using the top-slide, draw back the tool. Give it a touch with an oil-stone to get a good edge, and finish the boss and the recess at one cut. Advance the cross-slide till the tool just touches the boss, draw it back with the top-slide, put on about 0.004 in. of cut and traverse the length of the boss until you reach the radius; take care at this point or you will get chatter. Now, again with about 0.004 in. cut, traverse across the recess, and finally traverse across the side of the rim. Face the end of the boss to dimension.

Set a screw-cutting tool in the toolpost, and feed in (saddle still locked) to make the groove, As you get deeper, give an occasional traverse, a few thou one side and then the other, on the top-slide. This puts the cut on one side of the tool only, and gives a finer finish. The groove depth shown is for "Meccano" string; you can make it a bit deeper and wider with advantage.

Now for the hole. This must run true, or the wheel will wobble. If you have a Slocumbe drill that is $\frac{1}{8}$ in. dia. on the large part, centre with this, but go deeper than usual, to leave a short parallel part. If not, go in just as far as will allow a No. 31 drill to bed on the taper, not on the edge. Select a drill free from chips on the edge, and with lips of equal size. Give it a touch with a fine oilstone. Chuck in the drill

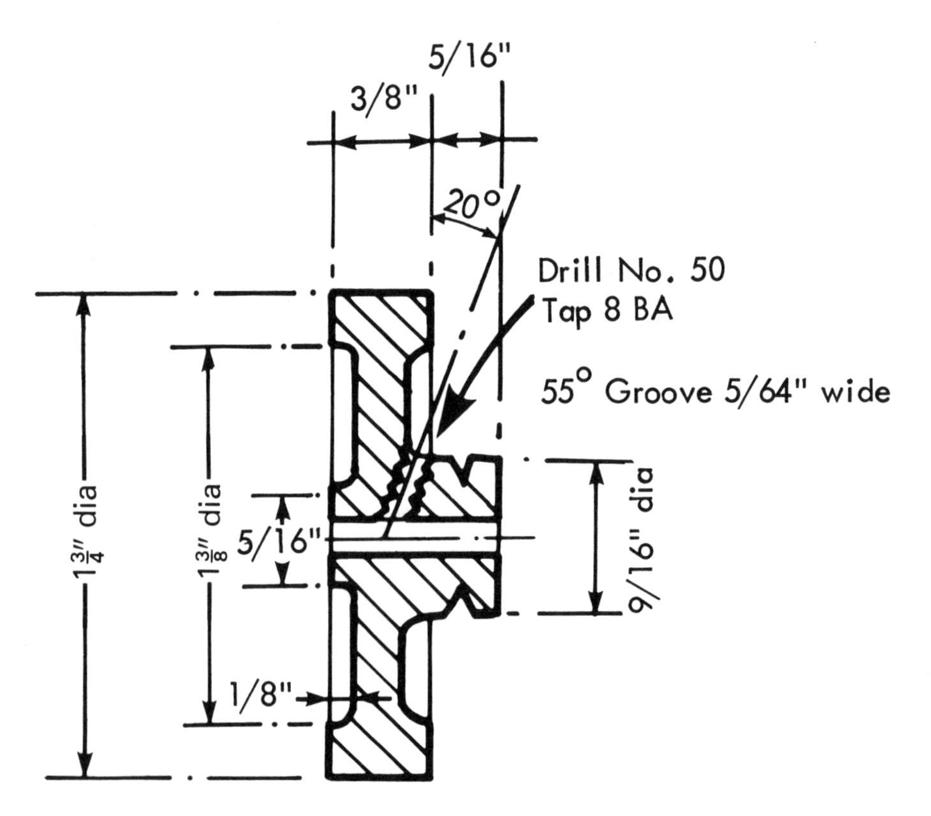

Flywheel—Brass Fig 2–10

chuck, but try several positions, until the point lines up best with the centre-hole in the wheel. Use the fast speed on the mandrel, and a slow and steady feed on the tailstock, but keep the drill cutting – don't let it just rub. Every so often withdraw the drill and clear the chips.Say about every $\frac{1}{8}$ in. at first, but at lesser intervals as you go deeper. When you are through far enough – say 1 in. – withdraw the drill, take it from the chuck, and push it into the hole. Spin the mandrel and see if it runs true. If so, chuck a $\frac{1}{8}$ in. reamer, and with a slow direct speed, push the reamer into the hole by sliding the tailstock.

If the drill wobbles, you have a problem. If the wobble is small, it may not be noticed on the running engine. Otherwise the best remedy is to bore out the hole to $\frac{1}{4}$ in. dia., and fit a bush. This can be made later, but is described now, in case I forget.

Chuck a piece of brass, $\frac{7}{8}$ in. long, and turn the outside diameter to about 15 thou larger than the hole in the wheel. Drill and ream a $\frac{1}{8}$ in. hole, as described above. Face the end. Reverse, and face the other end. Chuck a piece of $\frac{1}{2}$ in. steel, and turn down a length of $\frac{3}{4}$ in. to $\frac{1}{8}$ in. diameter, a nice push fit to the hole in the brass. Screw the end 5 BA about 5 threads. Face the shoulder, and then turn a short length to $\frac{7}{32}$ in. dia., or thereabouts. Push the brass onto this stub-mandrel, and hold with a 5 BA nut. Taking very light cuts, reduce the O.D. of the brass to a push fit in the wheel.

To return to the wheel, which is, by rights, still in the chuck. Part off as deep as you can, to make the rim $\frac{3}{8}$ in. wide, plus a few thou for facing. Take it out of the chuck, and go the rest of the way with a saw. Put a piece of shim steel between the saw and the face of the wheel, shaped like a letter U upside down, to prevent marking the parted-off face. Grip in the 3-jaw (outside jaws) by the rim, and with pieces of cigarette paper adjust it till it runs true. You may need paper both behind the rim and on its circumference to get it right. Beginners should note that even good 3-jaw chucks are only guaranteed to run true to 0.003 in., and that only when in one position, and using one of the keyholes. Don't expect miracles from them; precision chucks cost the earth.

Take a facing cut across the wheel, to remove the sawn stub, and bring to within a few thou of size. Machine out the recess, as before, but you have no boss to bother with this time. Lightly countersink the hole. Remove from the lathe, and centre-pop for the set-screw hole. Fit the bush, if one is needed. Grip the wheel in the drill vice, with pieces of paper to prevent scratches, and firmly prop it to the required angle. This is not of importance, so long as the drill clears the rim. Carefully set up by line of sight, so that the drill will hit the centre of the shaft hole. Drill No. 50. Tap 8 BA, and then poke a $\frac{1}{8}$ in. drill through the shaft-hole to clear the burr. Return the wheel to the lathe, and get busy with the Brasso, but don't polish the recesses – these ought to be painted a bright colour. Make a little set-screw by cutting the head off an 8 BA screw, and sawing a little nick in it; or you can use an Allen set-screw if you have one.

Cylinder Fig. 2–11

If you have a piece of thick-walled brass tube about the right size you can use this for the cylinder. Clean up the outside by spinning fast in the lathe and using fine emery. Cut off a piece $1\frac{3}{8}$ in. long, chuck lightly in the three-jaw, and face one end. Reverse in the chuck, face and bore out, using medium speed, light cuts and slow feed. About every third cut, traverse the tool without putting on any cut, to keep the bore parallel. Carry on until a $\frac{5}{16}$ in. reamer enters by about $\frac{1}{4}$ in. Don't ream right through.

If you have no such tube, you must drill and bore a solid piece, good quality brass preferred. Clean up the outside, cut off, and face as above, both ends. Grip in the chuck with about $\frac{1}{4}$ in. protruding, centre

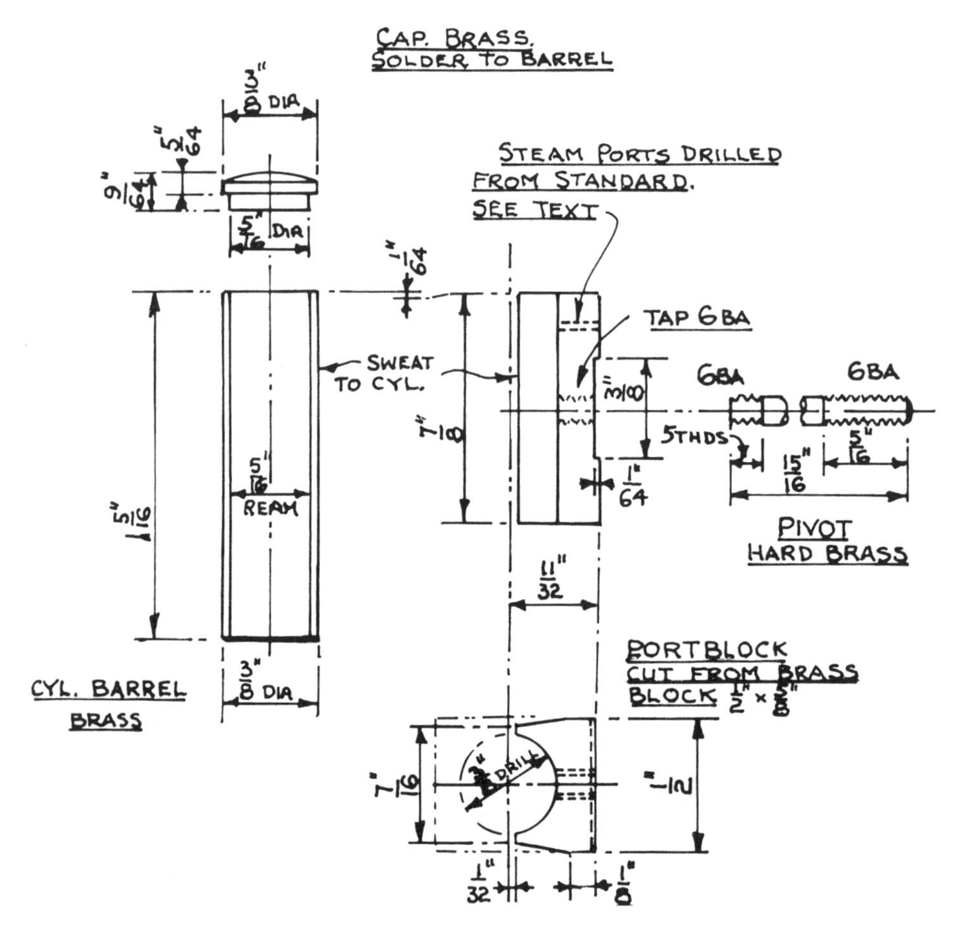

Cylinder Fig 2-11

with a Slocumbe drill, and follow with a letter "N" drill. Fastest speed, and slow feed on the tailstock. If the drill shows signs of running out, stop work. Grip a piece of $\frac{3}{8}$ in. square bar in the toolpost, and bring this into contact with the drill close to the work. Start drilling, and very gently advance the bar until the drill stops wobbling. Then remove it, and continue drilling.

The cylinder head is a simple turning job. Follow the procedure you used for making the safety-valve bush (without drilling it!) up to the parting off stage. The spigot should be a tight fit in the cylinder. Grip lightly in the chuck by this spigot, and round off the top till it looks nice. Polish with fine emery.

For the port block you will need a hefty chunk of brass – mine was chewed out of a piece of hexagon. Face off to $\frac{7}{8}$ in. long both ends, then cut to size. Smooth one side, as for the cylinder portface, making sure that it is square with the ends. Mark off a centre line on this face, and set up vertically in the machine vice on the drill so that both this line and the face are square, having first, of course, marked out for the $\frac{3}{8}$ in. hole. Drill carefully – don't force the feed, so that the drill will keep true. (Again, don't forget to have a piece of brass under the job for the drill to break out into.)

With dividers, mark out for the 6 BA hole in the portface. Set up in the drilling vice so that the face is dead horizontal – use a scribing block for this, if you have one. If not, chuck a centre-punch in the drill chuck, point down. Bring it down to the work until it just grips a piece of cigarette paper, and lock. Move the work under about the point using the paper as a sensitive feeler gauge. Drill No. 43 right through, and tap 6 BA. Mark out two lines, $\frac{1}{4}$ in. from each end of the block, and file the recess between them. The $\frac{1}{64}$ in. dimension is not critical, but it is good practice to try to get it right. Use a 15 thou feeler gauge and straight edge as a gauge. Now saw the block down the centre of the $\frac{3}{8}$ in. hole, and file to $\frac{5}{16}$ in. from the portface. Finish off the shape by filing the bevels, as shown on the drawing.

Offer up to the barrel, noting the $\frac{1}{64}$ in. overlap on the drawing, and mark round it. If the fit in the groove is too tight, ease slightly with a round file. Tin the hollow in the portface, and the barrel inside the marks. Apply flux, and clamp together with a piece of wood on the barrel, as shown in Fig. 2–12. Spread heat evenly from a small blowlamp until the solder flows. Allow to cool part way, and then set upright, leaving the clamp in place. With a small soldering iron, lightly tin the top of the cylinder, taking care no solder goes inside. If it does, poke a drill through and get rid of it. Cool and wash off all flux, first removing the clamps.

Put the lathe in middle back gear, and grip a $\frac{5}{16}$ in. reamer in the 3-jaw. Set the machine running and very slowly feed the cylinder over the reamer, easing back

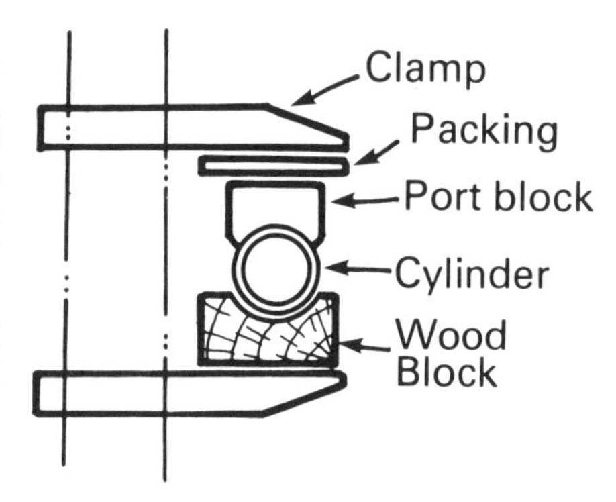

Solder Port Block to Cylinder Fig 2-12

frequently to clear chippings. Hold the job with a glove, not with the hands. (And certainly not with a piece of rag, which could get caught up in the chuck!) You *may* have to grip it with a cramp. You can do this job by hand almost as easily, of course.

Now tin the cylinder head, and when cold, press into the bore – the tinning will make it a tightish fit. Hold the block lightly in the smooth vice jaws, with the piece of wood over the barrel as used before, cylinder head on top. With a fine flame, heat the head *only* till the solder runs. Wipe off any surplus whilst still fluid, and allow to cool. Trim off odd lumps or splashes, and clean up with fine emery paper.

For the pivot pin, you need a piece of 0.110 in. silver steel – you can get this material in all BA sizes from stockists – otherwise you must turn down from a piece of $\frac{1}{8}$ in. stock. Using the tailstock dieholder, and taking more than usual care with the short thread, screw as shown. Start with the die wide open, and close it for a second cut, so that the thread is a sweet fit on a 6 BA nut. Screw the short end into the cylinder face, but not tightly at this stage. Find or wind a little spring, 24 s.w.g. bronze or 26 s.w.g. steel wire, 6 or 7 turns, about $\frac{9}{16}$ in. long when the ends have been ground flat. For beginners, bronze wire is easiest; just wind about 10 coils on to a $\frac{1}{8}$ in. steel rod, the coils nice and even, and cut off 1 coil more than the required length. Hold the rod against the grinding wheel so that it touches but not with any pressure, and with your fingernail advance the spring till the end coil is in contact. Apply pressure until the end coil is ground flat. Don't worry if the very end gets red-hot. Turn around, and repeat for the other end.

Piston Fig. 2–13

There are two alternative designs for the piston shown on the drawing. That made in one piece requires the use of a lathe, the other is intended for those who haven't got one. There is, of course, no reason why turners should not use the 3-piece design if they wish, but the single-piece design is more interesting! The method of making the piston etc. without a lathe is dealt with in the first chapter.

The drawing calls for stainless steel, but bronze will do, or even brass, provided the latter is a different sort from the cylinder. But free-cutting stainless is best. Now, this is a really nice turning job, one you can be proud of when you have finished. If you have a piece of $\frac{5}{16}$ in., it is worth trying to see if it will fit without machining. Chuck in the 3-jaw, and turn off the end where it is damaged from the original cutting. Cut a very small chamfer, and remove the sharp edge with finest emery, very lightly. A drop of thin oil in the bore first, then try it on the rod. If it won't go in at all, forget it. If it goes in but doesn't want to, that is ideal. If it slips in, again forget it; but if it slips in but wants to jump out again, you can use it. Whether or not, the first operations are the same, though you must use larger material if the above test on $\frac{5}{16}$ in. fails.

Cut off about $2\frac{3}{4}$ in., chuck in the 3-jaw, face and centre with a Slocumbe. Repeat for the other end. Remove the chuck from the mandrel, clean the hole for the centre, the threads and the register. Clean out the hole in your catchplate, and fit this to the mandrel nose. File a small flat on one end of the workpiece and fit a small carrier. Fit a soft centre to the headstock, clean out the hole in the tailstock, and fit a hard centre. Fill the hole in the free end of the workpiece, with tallow or hard grease, and swing between centres. Adjust the tailstock so that the work swings free, but does not rattle. Check this adjustment throughout the turning operation from time to time. Tie the carrier to the peg of the catchplate with a piece of string.

If the piece has to be turned down to fit the cylinder, do this first. Check after the first cut that you are turning parallel; if not, adjust the tailstock until it is correct. Now bring the diameter down to within 0.010 in. of the finished size, but not less than this. You must CUT stainless steel; scraping will tend to harden it. Change for a really sharp tool with a slightly rounded end, and with a feed of about 0.002 in./rev, bring the diameter down to 0.311 in, in one cut. Use plenty of cutting oil. If you are uncertain of your cross-slide index, then take a check cut over the first $\frac{1}{4}$ in. and measure it. Next, turn down to $\frac{1}{4}$ in. dia. for $1\frac{3}{8}$ in from the tailstock end, aiming at a good finish on the first $\frac{5}{8}$ in., which is the "big end" of the rod. From the shoulder so formed at the left-hand end, measure a distance of $\frac{27}{32}$ in. towards the tailstock, and with a sharp tool make a small groove; this is the measuring mark for the top of the big end. Between this mark and the shoulder, turn the rod down to the dimension shown. Go carefully, but don't take cuts less than 0.005 in. or the material may harden. Form a nice radius at each end.

At $\frac{9}{16}$ in. to the left of the shoulder, make a similar mark. This is the top of the piston. With a screw-cutting tool, form the two small grooves – feed in about 20 thou – and then the bevel on the top of the piston.

Return to the tailstock end, and $\frac{9}{32}$ in. to the right of this shoulder partially part off – leave a piece $\frac{1}{8}$ in. dia. uncut for the present. Return to the piston end, and carefully partially part off just to the left of the groove, i.e. at the top of the piston. Remove from the machine and with a fine saw cut off the waste. Remove the catch-plate, and making sure threads and register are clean, refit the 3-jaw. Grip the piston lightly, head outwards, and face the end. Form the bevel. Reverse in the chuck, still gripping by the piston. Lock the headstock mandrel with the back gear. Set up a scribing block on the lathe bed, and adjust to the top of the $\frac{1}{8}$ in. pip on the big end. Scribe a line on both sides of the work at this height. Repeat, using the bottom of the pip. You now have reference lines to file to in forming the flat of the big end.

Reverse in the chuck, and grip by the big end boss. Spin fairly fast, and remove any burrs from the piston. Clean off all dust, apply a spot of oil, and try on the cylinder. It should slide right to the top, and compressing the air in front, try to pop back again. If it slides in but is tight, it must be lapped to fit. Set the lathe in lowest back gear. Apply a good dollop of thick metal polish to the piston, set the machine running, and with a steady up-and-down motion, slide the cylinder back and forth on the piston. Don't let it stay in one place. Every now and again, slide the cylinder off, wash in clean paraffin, apply a little oil, and test for fit. Don't leave the two parts together whilst you go for a cup of tea; the cylinder will shrink onto the piston, and won't come off! When you have finished, the piston will seem to rattle a little when it is dry, but fit beautifully when oiled. This is as it should be. Finish off with a thorough cleaning out of the cylinder, and the grooves of the piston.

Remove the work from the lathe, grip in soft jaws of the vice, and file the flats on the big end. Get a smooth finish, and then remove the pip on the end with a fine file.

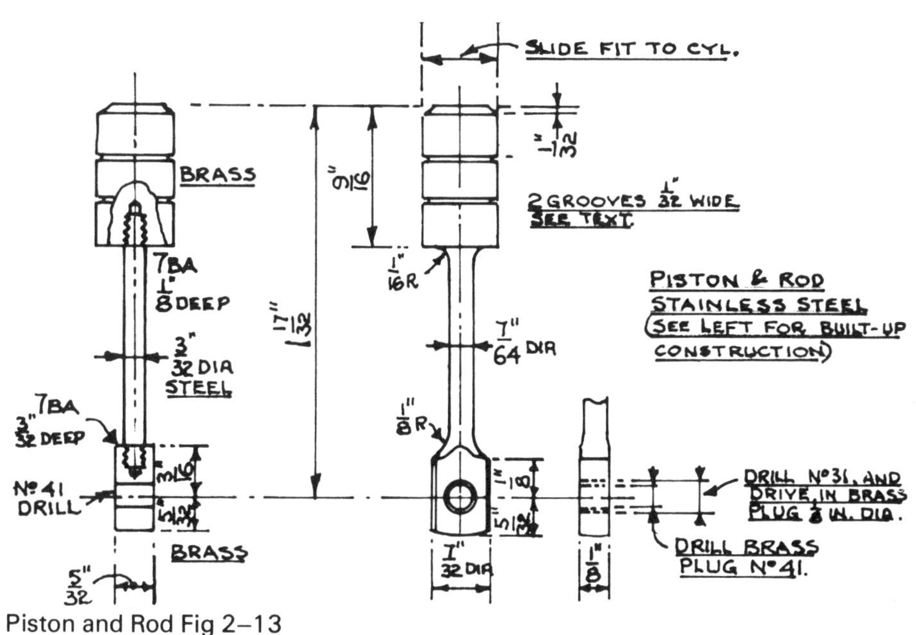

Piston and Rod Fig 2–13

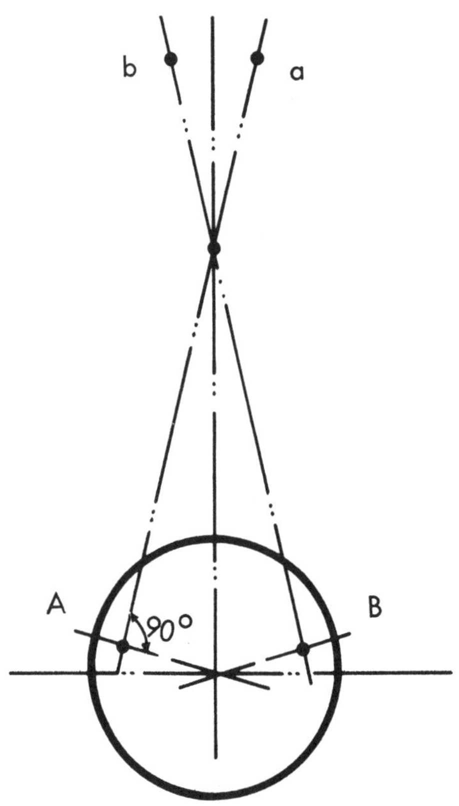

Fig 2–14 Crank position for drilling cylinder ports

Chuck by the piston lightly, and set with a square on the lathe bed so that the flats are vertical. Set a scribing block to exact centre height, and scribe a line along the flats of the big end. This is the centreline for the bearing. Clamp in a vee-block with the big end up in the air, and the piston head flat on the bed. Set the scribing block to $1\frac{17}{32}$ in., and scribe a line across the flats. Lightly centre-pop at the cross lines. With a small centre-drill, deepen the pop mark, and then drill No. 31. Broach very slightly taper – a couple of twists will do. If you have no broach, twirl a tapered round Swiss file in the hole. Press in a piece of $\frac{1}{8}$ in. brass wire, and file flush with the flats, but don't file out the scribed lines. Find the centre of the plug using these lines, centre-pop, and drill No. 41. Polish out the scribed lines, and then lightly countersink the hole each side, just to remove the burr.

Drilling the Cylinder Ports

Attach the boiler to the firebox, and then fit the standard. Oil the crankshaft and fit to the bearings. Attach the flywheel, but don't tighten the screw more than a pinch. Oil the piston and fit to the cylinder. Assemble to crankpin and standard, fitting the spring, washer and nut. Set the crank to position A in Fig. 2–14. This is looking *at* the cylinder side of the engine. With a No. 53 drill held in a pin-vice, mark through the hole "a" in the standard, twirling the vice round several times to make a good mark. Set the crank to "B" and repeat through hole "b". Make sure that nothing slips whilst doing this – use a clamp if need be – and hold the cylinder firmly against the portface.

Remove the cylinder and take out the piston. Rest the cylinder, port upwards, in a small vee-block, and clamp with the portface level. Line up and drill each port No. 53, right through. Do not remove burrs from the portface with a drill or anything like that. We shall lap them off in a minute or so. Reassemble on the standard, and peer through the ports as you rotate the wheel. You should see each hole open fully when the crank is at A or B, neither hole open at all when the crank is at dead centre, but each hole beginning to appear as the crank is rocked each side of top centre. It doesn't matter if the exhaust port is a little before dead centre when it closes; but if the steam port is not opening just after dead centre (going the right way round) you can enlarge it to No. 52, using the drill in a pin-vice. Enlarge the one in the cylinder first, and if that doesn't quite do it, repeat on the standard. Fortunately, you are not likely to get both ports open at once on dead centre, unless you have made a real error in dimensioning the standard.

Remove the cylinder and crank, and detach the standard. Refit the cylinder to the standard, but without the spring. Using metal polish, lap the portfaces together, using an oscillating motion. Part the two faces frequently, and don't let them get dry. Towards the end, add a little oil to the metal polish. You should finish with a nice surface over both ports and similarly over part of the lower face. Wash well with paraffin after polishing the bright parts.

Initial Erection

Fit a lead weight in the base – this is to stop the engine waltzing about when running, and to make it less top-heavy. Touch up any scratches on the paint, and polish out any similar in the brightwork. Assemble the standard, for good this time; get the bolts tight. Insert the shaft in the bearings with a trace of oil, and adjust until it runs freely. You will have to bend the lugs a little, in and out or up and down, and it is worth taking some pains over. Fit the flywheel towards the end of this operation, with about $\frac{1}{64}$ in. side-play.

To make the steam pipe, straighten a length of $\frac{1}{16}$ in. copper wire by holding one end in the vice and giving a good heave on the other end with a pair of pliers. Cut off about 7 in., and offer it up to the engine, bending it until you get the curves to suit your fancy. Make a mark on it where it enters the boiler, and cut it off about $\frac{1}{2}$ in. longer. Use this to measure off a length of $\frac{1}{8}$ in. copper pipe. It should not need softening for this job unless it has been bent about a lot. File one end square, and ream out the hole in the end. This is the cylinder end. Bend the pipe to match your wire template. File the other end square and ream out the hole, Blow through it with a tyre pump to remove any debris. With a taper pin, enlarge the cylinder end of the pipe till it is a tight fit in the hole. Put a clothes peg or a paper clip on the other end to stop it from falling into the boiler. Put a little taper pin, slightly oily, into the exhaust port, to stop solder running into it, and solder the pipe to the portface. You will find resin cored solder best for this joint, provided all is clean. Remove the clothes peg, and reposition the pipe till it looks well, then solder to the boiler. Check that no solder has run into either port. If it has, remove it with a 53 drill held in a pin-vice.

You will find that the standard has stiffened up a lot now, and will almost certainly have to readjust the bearings. Assemble the cylinder after this, and adjust the vertical alignment of the standard if necessary so that there is negligible difference between the space between big end and the crank flange on either dead centre. Use washers to control sideplay if need be. Put a drop of cylinder oil on the piston and portfaces; she should now spin freely, with a trace of a "puff" from the exhaust.

Safety Valve Fig. 2–15

You can use any "LBSC" design safety valve, but screwed $\frac{1}{4}$ in. x 32; I don't like 40 t.p.i. on jobs that must repeatedly be unscrewed. Or, as already mentioned, you can buy one ready-made. That shown on the drawing is, however, quite easy to make. Chuck a piece of $\frac{3}{8}$ in. hexagon brass, face the end, and turn down to $\frac{1}{4}$ in. for $\frac{7}{32}$ in. long. Get a good finish on the face of the shoulder. Recess the corner with the point of the knife tool. Use the tailstock die holder, and screw $\frac{1}{4}$ in. x 32 t.p.i. Bevel the corners of the hexagon. Part off in. to the left of the shoulder. Chuck a piece of scrap brass about $\frac{1}{2}$ in. dia., centre, and drill $\frac{7}{32}$ in. Tap $\frac{1}{4}$ in. x 32 and countersink the hole. Screw the part-finished valve into this holder, and face the end to dimension. Centre, and drill right through No. 34. Follow with a No. 6 or $\frac{13}{64}$ drill until the corner of the drill enters about $\frac{1}{32}$ in. Use a D-bit or a flat countersink cutter – or, at a pinch, a $\frac{3}{16}$ in. milling cutter – to flat-bottom the hole. Ream the small hole $\frac{1}{8}$ in.

Machine the outside to any fancy shape you please, but leave about $\frac{3}{32}$ in. of hexagon. Find a $\frac{3}{16}$ in. hard steel ball, and give this a tap with a hammer so as to form the seat inside the recess. Remove from the holder, and put the screwed adaptor aside for use in the future.

Chuck a $\frac{3}{16}$ in. bronze ball in the 3-jaw with a shade projecting, and with a very sharp tool and at top speed, make a tiny flat face, about $\frac{1}{16}$ in. at most in diameter. Centre lightly with your smallest Slocumbe drill, and follow with No. 54, right through. Tap 10m BA – hold the tap in a pin-vice and be very careful! Remove from the chuck, and put it in a little box where it can't get lost. Chuck a piece of $\frac{1}{16}$ in. hard brass wire 1 in. long, and screw $\frac{3}{16}$ in. at one end, $\frac{5}{16}$ in. at the other, 10 BA. Grip in the tailstock chuck, $\frac{3}{16}$ in. screw projecting, screw on the ball, flat to the right, as far as you can. Bring up to the headstock, chuck the ball lightly, and screw on about $\frac{1}{2}$ to 1 turn more.

Safety Valve Fig 2-15

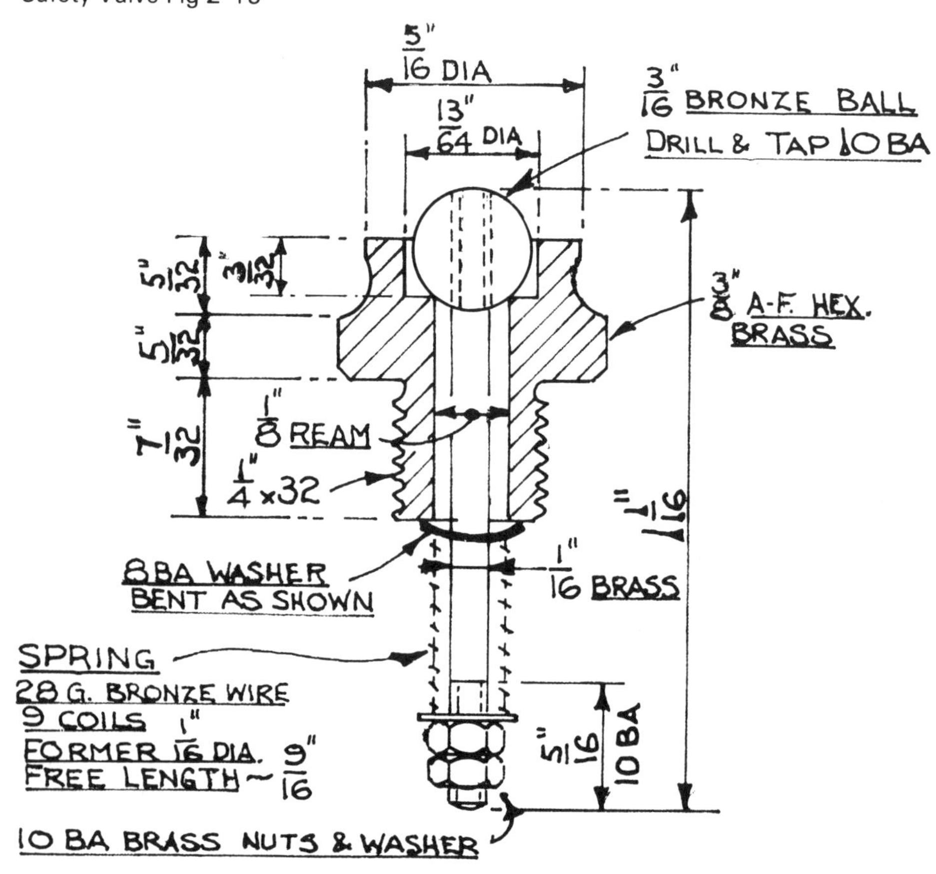

Wind a little spring of bronze wire, about 9 turns of 28 s.w.g. wire on a $\frac{1}{16}$ in. former, in the same way as for the cylinder. Assemble the valve with a 6 BA nut. Get the kitchen scales into the shop, and adjust the nut until it requires a force of between 2 and 3 ounces to open; you can find this accurately by putting a drop of meths round the ball. It will run down when the valve opens. You may have to experiment a bit with the number of coils for this job. The pressure will be around 10 p.s.i. at this setting. When adjusted, fit a locknut. Screw into the boiler with a fibre or leather washer; don't use aluminium, as these tend to corrode.

Lamp Fig. 2–16

First find your wick! This is hard to come by, and I have used that from the little "kelly" nightlights that are still around in the shops. It is $\frac{3}{16}$ in. dia., and three wicks give a reasonable steaming rate. $\frac{1}{4}$ in. might be better, but if larger, reduce the number to two.

The tank is made from shim-brass, 0.010 in. thick, though 0.005 in. would do for the barrel. Flange the ends on a former, exactly as for the boiler. Drill the three $\frac{3}{16}$ in. holes for the wick tubes and the $\frac{1}{8}$ in. air vent in the middle of the top plate, using the former as a support. Remove the burrs. Cut a piece of shim 1 in. wide and $4\frac{15}{16}$ in.

Lamp (not to scale) Fig 2-16

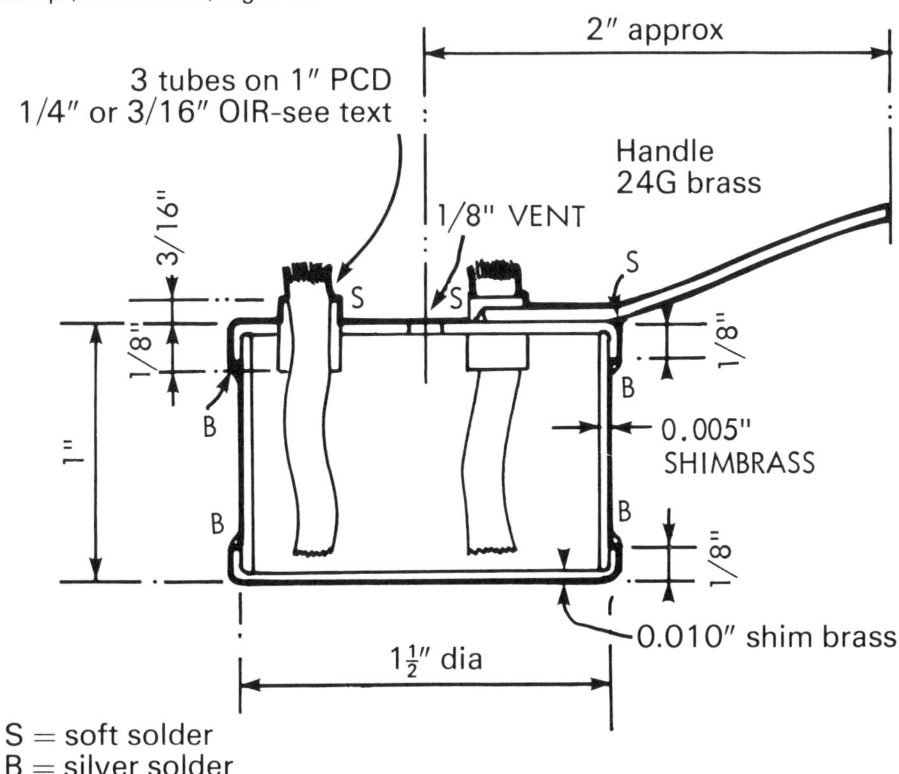

long; roll around a piece of $1\frac{1}{2}$ in. stuff and manipulate it until it will fit nicely inside the flanged ends, with a close overlapped joint at the side. Make a scriber-mark down this vertical joint, and take apart. Flux the very ends of the sheet, and clamp together with a small *steel* clamp so that the scribed line mates with the sheet end. Silver solder the joint, but go easy with the heat, or you may find the whole issue suddenly collapsing into a pool of molten brass! Adjust your blowlamp to give a very soft flame. Cool to black, and quench. Scrub with a wire brush, and restore the shape. Fit to the endplates, fluxing the flanges and the little cylinder (including the existing joint). Hold together with a clamp, or twist a bit of black soft iron wire round to hold the parts together. Braze the two flanges. Cool to black, and pickle.

To make the wick tubes, cut a strip of 0.005 in. brass shim about 2 in. long and soften it. Clean off the oxide, and cut three pieces $\frac{5}{8}$ in. long. Roll these round a piece of $\frac{3}{16}$ in. rod. It will help if you "preform" the end $\frac{1}{16}$ in. with a pair of pliers. Then cut out the handle from a piece of thin brass sheet – about 28 s.w.g. Shape to choice! Remove the body from the pickle and wash well. Scrub with a brass wire brush and give it a good clean up, removing odd lumps of silver solder if there are any. Enter the wick tubes till $\frac{3}{16}$ in. sticks out, and push in a slightly tapered piece of $\frac{3}{16}$ in. wooden dowel. This will open them to a good fit, and hold them in place. Soft solder them in, with a good fillet, using tinmans soft solder, not resin cored stuff. Then remove the pegs and solder on the handle, so that the seam in the barrel is on the opposite side. Remove flux, and polish. Then cut three pieces of wick sufficently long to reach the bottom of the tank and stick out by something over $\frac{1}{8}$ in.

Test Run

Fill the boiler with clean water, empty out, and measure the quantity. Take two-thirds of this as a normal filling. Use hot water, to save spirit. Put a drop of oil – 30 SAE – on each bearing, and cylinder oil on the portface and up the cylinder. Fill the lamp through one of the wick-holes, about $\frac{7}{8}$ full. Tighten the safety valve with a spanner, not too tight; in fact, if you have a leather washer, finger-tight will do. Light the lamp, and if necessary trim the wicks. Then into the firehole with it.

After a while, the boiler will start to sing, and you should gently revolve the flywheel. Dollops of water will emerge from the portface and the exhaust, so make sure no tools are around. It may take a little while for the whole to warm up and/or sufficient pressure to generate to drive the new engine, so don't be impatient. Cock an eye on all joints, and make sure there are no leaks under pressure. As soon as the engine is buzzing merrily, blow the lamp out; fetch the family; and light up again. They will be as pleased as you are! Finally, before the spirit is all used up, stop the engine and set with exhaust port open. Continue firing, and make sure that the safety valve can cope. If she doesn't blow off within 2 minutes of stopping the engine, the spring is too tight. There should be no blow when running, and start to blow within 1 minute. If she does not reseat, give a gentle tap with a hammer, but you should expect a *slight* wisp.

When the lamp starts to go out, remove it and dowse the wicks. Let the engine cool, and pour out the water. If this is less than half you put in, well enough. If more, reduce the quantity specified to the engine-driver. If there is none, then either you have a leak, or the lamp was too full!

Test-run the engine several times, and apply pressure this way and that to the engine standard. If you find such pressure causes the engine to run faster, make the shift permanent. When the whole gets hot, particularly the steam pipe, some distortion always occurs.

Finally, give the engine a good clean.

Get a supply of proper cylinder oil, and put some in a small bottle. Trot up to the hospital, and see if you can scrounge one of the disposable hypodermics they use (without the needle) which has only had water in it, for use as a lubricator (or buy a new one – they are cheap enough). Write out a set of working instructions. Find a nice box, and some packing. Then take it along to your favourite nephew or niece, 5 years old or upwards (up to 50!) and watch his eyes light up when he sees what it is. (Don't forget to take some meths with you, by the way; such a waste to have to run it on Vodka!)

That is all there is; simple, isn't it?

Chapter Three

ELIZABETH

A HORIZONTAL STEAM POWER PLANT.

The previous engine was of a type which has given enjoyment to children of all ages for over 100 years, the vertical "engine" mounted on the vertical boiler being about the simplest arrangement that could be conceived. For the next one, however, I have designed a separate boiler and engine mounting. Opinions differ as to which layout is the more attractive, but there is little doubt that the horizontal arrangement offers more scope for individual builder's originality, provides the opportunity for greater variety in the detail, and is probably better adapted for driving small models. The baseplate, for example, shown in the photograph as a simple piece of wood, can be of metal drilled to suit "Meccano" models (4.1mm or No. 20 at half-inch pitch) extended to carry a countershaft and some of the little tinplate model machines that are now available, or even made up as a floor to an engine-room built round it.

The engine will run for rather longer than will the vertical and though the cylinder is a shade smaller, will develop about the same power. The boiler is, however, arranged for firing using solid fuel of the "META" tablet kind. This is more convenient, and you can arrange for short runs just to show off the engine quite simply; just load up with half the number of tablets. However, if there are any tiny tots in the family please be careful and store these fuel tablets out of their reach. They look like sweeties and don't taste unpleasant, but they ARE poisonous if taken in any quantity. (The substance is, in fact, the active ingredient of many slug baits used by gardeners.) Don't be put off by this – there are literally tens of thousands of META-fuelled appliances in use, especially for picnic stoves; just take the same care as you would with aspirin.

The engine shown in the heading photograph was made almost entirely from odds and ends I had lying around, and the drawings describe it "as built". The engine standard, for example, is made up from a piece of brass angle left over from making a frame to carry some floor tiles, and is mounted on two pillars which came from an ancient wireless set I had just dismantled. You can modify this to suit what YOU have available as you please; the only thing you must take care over is that the holes carrying the cylinder and crank are in the same positions as shown, and that the crankshaft bearing is the same size. The safety valve and steam cock can be bought; it's hardly worth the time to make the cock, and they can be had from Stuart Turner Ltd of Henley-on-Thames, or from A. J. Reeves & Co Ltd, Holly Lane, Marston Green, Birmingham. Commercial safety valves, though, are usually supplied to blow off at rather a higher pressure than you need, so I have given details of the

Elizabeth

construction of this one. Copper tube for the boiler, steam and exhaust pipe, and brass or copper sheet can be had from the same people. (Or see the advertisers in "Model Engineer" or "Model Maker")

Now, I am asked often enough whether one can use brass tube for boilers etc. The answer is "Yes, if you like". Most of the similar engines sold when I was a lad had brass boilers. See page 9, Ch 1. One other point here. Both copper and brass tube are likely to be offered in metric sizes now. This doesn't matter at all. You can use anything from 40mm to 50mm diameter (just alter the housing to give the same fire-space round the shell) and from 0.6 to 0.8mm thick for the tube and from 0.8 to 1.0mm thick for the ends – it will be quite safe! Both $\frac{1}{8}$ in. and $\frac{5}{32}$ in. diameter (outside) steam pipe should still be available for many years yet, but if not, use $3\frac{1}{2}$ or 4mm, 0.5 or 0.6mm thick, preferably the former.

For the cylinder I was fortunate enough to find a piece of $\frac{1}{4}$ inch bore brass tube which exactly fitted a stock piece of brass bar; a bit tight, but after lapping with metal polish it slid back and forth very easily. You may not be so lucky, in which case you have two alternatives – make the tube fit the piston material, or vice versa. If the tube is too small (assuming you are using a lathe) turn the piston material down to suit after putting a reamer through the bore. Note; *Don't* grip too tightly in the chuck when reaming. If your only piston material is a very slack fit in the bore, then you must either turn down from a much larger piece of stock, or use the alternative design of packed piston shown in the drawing. If you have no lathe, then I'm afraid you will need to take more time, as described on page 11. In what follows I shall assume that you have a lathe available.

Engine

Start by making the engine part – Fig. 3–1 Then you can test this with compressed air or a tyre-pump before building the rest. Saw off the correct length of $\frac{1}{4}$in. bore tube and file or turn the ends square. Ream the bore as mentioned above if need be, otherwise just spin it at a moderately high speed and polish with very fine emery cloth wound round a stick. For the piston, brass, stainless steel, or german silver will do, but NOT aluminium or ordinary steel, both of which will corrode in contact with the brass cylinder and water. Set in the 3-jaw chuck and face the end. Drill about $\frac{3}{4}$ in. deep with No. 45 drill, and tap 7 BA, guiding the shank of the tap with your tailstock chuck to keep it straight. If the piston-to-cylinder fit is satisfactory as it is, make the three $\frac{1}{32}$ in. grooves with the point of a screwcutting tool; otherwise use a $\frac{1}{16}$ in. parting tool to make the grooves for the soft packing shown as an alternative. Carefully remove any burrs. Part off to length. In the absence of a lathe, proceed as described in the section at the beginning of the book. The main thing in this case is to get the hole upright – it won't matter if it isn't quite in the dead centre of the piston.

The piston rod can be of brass if you have no stainless steel of the right size. It is very important to avoid getting a "drunken" thread, and the usual cause of such is attempting to make the thread with the die too close. It is better to rough out the thread with the die well open, and then take a second pass with the die readjusted – check it against a commercially made screw. It is better to have the thread on the rod over rather than undersize, as you want it to be a tight fit in the piston and bearing end. The latter is of brass, though german silver will do, and gunmetal better than either. *Don't* try to drill and tap a piece of stuff which is only $\frac{1}{8}$ in. thick; Use a thicker piece - say $\frac{1}{4}$ in. square. Drill and tap, then assemble temporarily to the piston rod and file to thickness. This enables you to check that the faces of the bearing are parallel to the line of the rod. The $\frac{1}{8}$ in. dimension is not critical. Having done this, assemble the piston as well, measure the $1\frac{17}{32}$ in. dimension from the piston head to the hole, and drill the latter, taking care that all is held square whilst you do it. The hole should be an easy fit on $\frac{3}{32}$ in., and if your drills are "tired" and hence unlikely to cut to size, use No. 42 or 2.3mm instead. This done, check that the hole is square by putting a drill through it, or a piece of rod, and insert the piston into the cylinder. Check with a square and, if need be, *carefully* bend the rod a trifle. When satisfied you can put a small blob of solder on the screwed joints if you are not absolutely sure they are dead tight. (You can use Loctite on the threads instead if you have any handy)

The port block comes next. This looks difficult, but only needs a little care. In the past many constructors have advocated

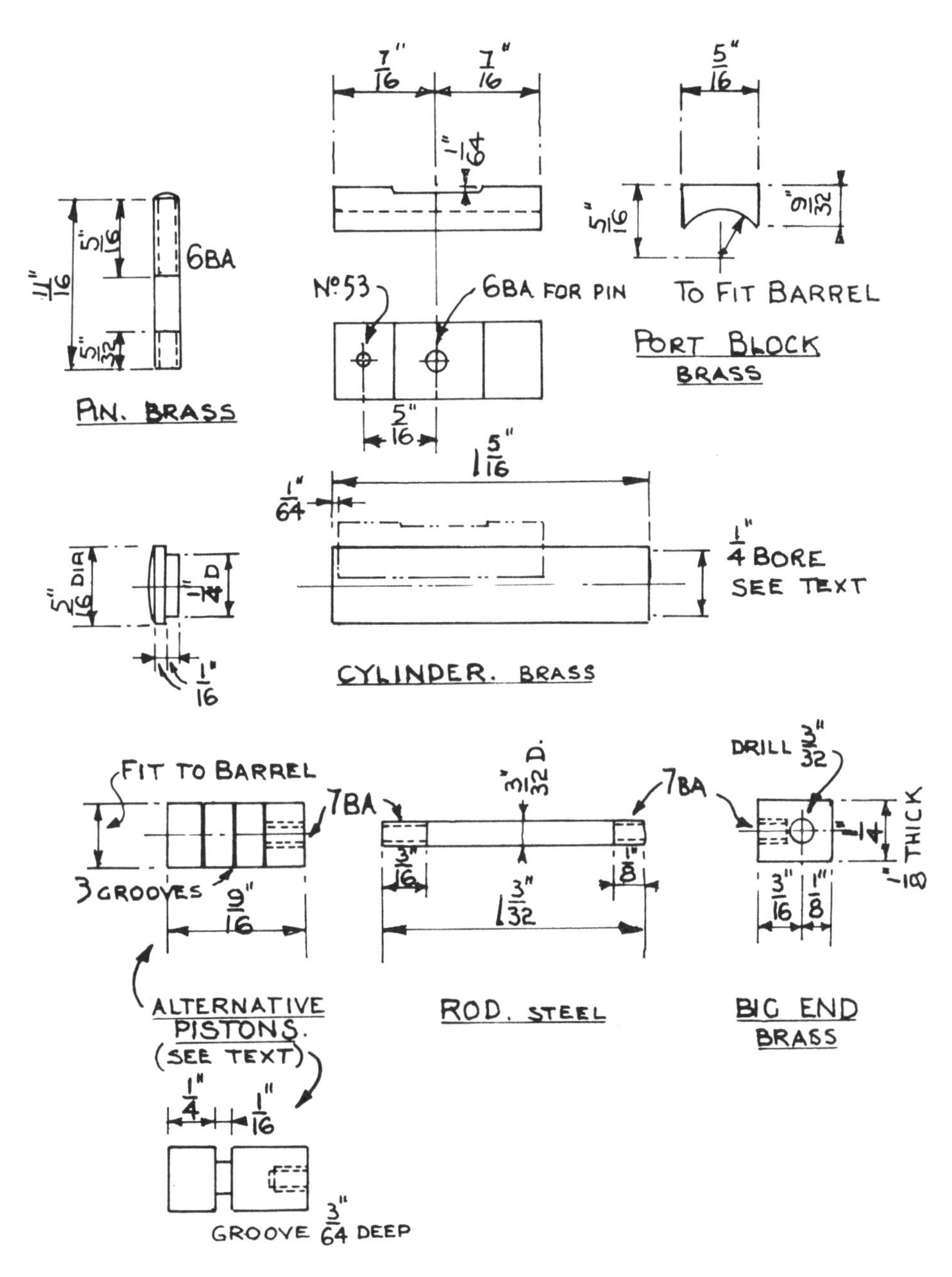

Cylinder Set Fig 3-1

turning the semicircular groove on an angle-plate in the lathe, but it will serve just as well if you file it. Or you can use the method described on page 37 for the first engine – drill down a block of brass and then cut to shape. However, filing makes a change! Use marking blue to see that it fits reasonably well, especially at the end with the 1.5mm hole in it. This is the steam passage. When the groove fits tolerably well, file the flat face till it is as near as you can measure parallel to the groove face – we shall get it dead right later on. Mark out for and drill the two holes, and tap the centre one 6 BA. Set this aside, and make the little plug which forms the cylinder head. The spigot shown is merely to hold it in place whilst soldering, and to reduce the cylinder volume a little, that's all, and you can work accordingly. You must now solder the parts together, with the end of the port block flush with the end of the barrel of the cylinder, (the small hole at the head end, of course) so the cap will protrude beyond it. Clean the faces free of grease and give them a rub with fine (oil free) emery, or perhaps better, very fine sandpaper. Now, I use solder paste for this sort of job, but many people don't seem to have heard of it. This is a mixture of solder powder and flux, and the brand I use is called "FRYOLUX" (there may be others). If you can't get it at the local ironmongers or D.I.Y. shop, then try K. R. Whiston, New Mills, Nr. Stockport, Cheshire. Apply this in an even, but sparing, coat both to the port block and the cylinder barrel and either lightly clamp or bind it in place with iron wire. Fit the cylinder head and apply a little paste round the joint. In all this take care that none of the paste gets inside the cylinder. Have a stick of solder, the ordinary tinmans sort, handy. Heat gently with your blowlamp, or even a spirit lamp; as soon as you see the solder in the paste melt and run, apply just a little more solder from the stick to make a nice fillet. Not too much, mind you, or you will have to file it off, and that sort of "Brummagen" method isn't "proper"!

If you can't get this stuff, then use the good old-fashioned way – it is, in fact, better, but takes a lot more care. Tin both surfaces to be soldered separately; wipe on some flux, get it hot, apply solder either in wire form or from a soldering iron, and clean off with a dry rag whilst still molten. This leaves a smooth, silvery surface. Clamp or bind together as before, heat again, and apply a little solder from the coil. (The 16 gauge solder wire is best for this sort of job, rather than the tinman's standard stick) This will flow into the joint by capillary action, mating with the already tinned surfaces to make a sound joint. After which – whichever method you have used – clean off all flux, and I recommend that you boil the assembly in a saucepan to make sure all is removed.

We must now resurface the flat of the portface to make sure it is parallel to the bore. Set on a flat plate, and with a surface gauge or even a steel rule, check that the top of the cylinder lies parallel to the surface of the plate. File till correct, at the same time keeping an eye on the $\frac{5}{16}$ in. dimension to the cylinder centreline. When all this is in order, file across the face to make the centre recess – this need only be a few thousandths of an inch down; it is there only to ensure that the cylinder beds just at the ends. However, I always make it about $\frac{1}{64}$ in. down to be sure. Now make the brass pivot pin; the same remarks about the thread apply here as to the piston rod, and if you have a lathe it's best to use the tailstock die-holder. Note carefully; the $\frac{5}{32}$ in. screwed end must be reduced by trial until the pin screws into the port block without bottoming, to avoid denting the wall of the cylinder. You may find that some solder has got into the hole, and if so you will have to clear it with your bottoming (or plug) top. Fit to the hole and check that the pin stands square. If it doesn't, "bend it straight", but take care in

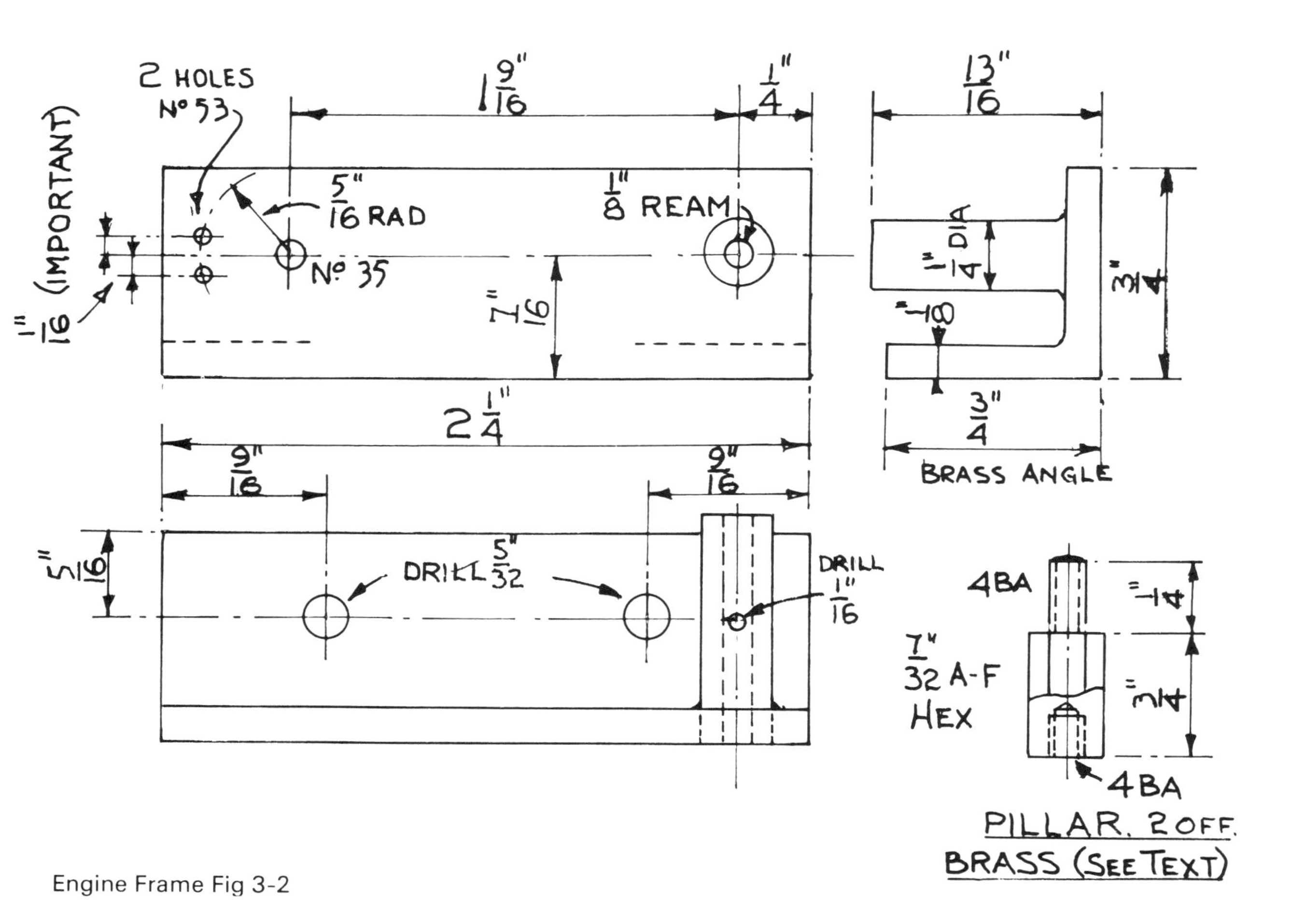

Engine Frame Fig 3-2

doing so. If it is a long way out of square you may have to start again, I'm afraid, even making a new port block if it's the hole that is awry. If you haven't a piece of 0.110 in. stock from which to make the pin, use a 6 BA screw and fill up the threads in the middle with solder; this will serve until you can get some. Don't fit the pin permanently yet. Finally, carefully drill through the No. 53 hole into the cylinder barrel.

Engine Frame

Now for the engine frame, Fig. 3–2. You may wonder why we can't use a substantial piece of $\frac{3}{4}$ in. square stuff for this. The answer is that it would so chill the cylinder that all the steam would condense and you would get no work out of the engine even if it ran at all; and by the time the block had heated up the boiler would be empty! Pieces of brass angle are found in most workshops, but failing this there is no reason why you should not make up a frame by brazing a couple of lugs onto a piece of flat. Indeed, you can improvise or redesign to your heart's content; the only dimensions which matter are those of the centre-distances and sizes of holes on the vertical face and the proportions of the bearing. So, having made this adjustment, or found your piece of stock, proceed as follows.

Clean up the face designated as vertical and very lightly mark the centrelines of the crank bearing and the 2.8mm pivot hole. Lightly centrepop the latter, and set your dividers carefully to $\frac{5}{16}$ in. radius; with them draw the little arc on which the ports will lie. VERY lightly pop where this arc crosses the previous centreline. Now use an eyeglass and set your dividers as carefully as you can to $\frac{1}{16}$ in. radius (1.6mm) then use these to mark out for the ports themselves. To check, reset the dividers at $\frac{1}{8}$ in. (3.2mm) and examine the intersection of the scribed lines against the point of the dividers. Very lightly indeed centrepop these intersections, having first sharpened your centrepunch. (A tap on the head of your scriber may be better to make a tiny pilot pop) Examine these pops with your eyeglass to check, and when satisfied, deepen them (I use a diamond shaped bit in my archimedian fretwork drill for this sort of job; one can place the point exactly under magnification, after which slight pressure and slow rotation of the drill will form a small indent. This can be moved sideways if need be, and then deepened as needed. No centrepopping is required)

Reset the dividers to $1\frac{9}{16}$ in. and mark out for the bearing hole. (This, be it noted, will be $\frac{1}{4}$ in. dia. to accept the bushing which is soldered to the frame). Draw a $\frac{1}{4}$ in. circle at this point, so that you can see if the drill wanders. However, drill the two No. 53 (1.5mm) holes first; these are important, and if they go wrong you won't waste time over other work, but start again – or plug them and redrill. The other holes may then be made, but the 2.8mm one should be drilled last, undersize at first and then opened up. It needs to be a close running fit on the cylinder pivot pin. For the crank bearing bush, if you have a lathe you can drill and ream it there; if not, solder in a piece of $\frac{1}{4}$ in dia. brass bar, mark out from the centre of the pivot pin and drill and ream $\frac{1}{8}$ in. In the absence of a $\frac{1}{8}$ in. reamer, drill No. 31 and then follow with a $\frac{1}{8}$ in. drill feeding slowly. Not shown on the drawing, I'm sorry, should be a $\frac{1}{16}$ in. countersunk oil hole in the middle of the length of this bush. Drill the two 4 BA clearance holes (No. 26 or 4mm) for the standards.

File down any projection of the bush and then flat the whole port surface using new fine emery on a flat plate or piece of glass. The part that really matters is shown shaded on the drawing, but it is easier to do the whole. Now check the bedding of the cylinder portface. Push the spindle through the hole and with gentle pressure rotate the assembly. You will see where it touches; if even all over, well and good. If not, then you must ease the high spots, using a fine file if the difference is large, or a scraper (or the end of a newly sharpened penknife if you haven't one) if the faces almost but not quite bed. It is important that a good bed is obtained adjacent to the ports, but unless the faces are badly out a certain amount of bedding in will occur during running.

Crankshaft

Make the crankshaft next – Fig. 3-3. The shaft can be of silver steel but you will almost certainly have to polish it; I know it is "ground" but the surface finish isn't very good for all that. Polish till it is a very easy running fit in the bush. Screw the end, again taking care that you don't get a drunken thread – I needn't repeat myself over that! The crank disc is shown as "balanced" but there really isn't any need for this – it can be a plain disc of steel $\frac{7}{8}$ in. (or even 1 in.) diameter with no loss. The important point here is that the holes must be drilled parallel, and that the 5 BA hole should be tapped truly upright. The only recipe here is "take care", and use your drilling machine, or the drill stand with your portable drill, and make sure that the disc is held dead flat in the vice. The small hole for the crankpin is drilled No. 43 and the $\frac{3}{32}$ in. pin very slightly tapered at the end and then driven in. (You can, of course, drill to size and use Loctite retaining compound if you like)

Now, you *can* make the crankshaft of $\frac{5}{32}$ in. steel if you wish, provided you drill the bearing bush etc. to suit. You can then fit Meccano pulleys and the like direct onto the shaft. The end will then be screwed either $\frac{5}{32}$ in. whit (same as Meccano bolts) or 3 BA, though the latter might need a bit of Loctite or Araldite to retain it, as 3 BA is a bit slack on this diameter. I worked the other way round – used $\frac{1}{8}$ in. for the spindle to reduce friction, and then bushed the Meccano pulleys which my daughter wanted to use with the engine. In the event the engine has been used only with the smallest of the Meccano pulleys, as its operating speed is high.

Flywheel

Now for the flywheel. (Fig 3–3). No problem if you have a lathe. You will note that I have given no dimensions for the recesses in the sides – this is very much a matter of taste and does depend a bit on what tools you have set up in the turret. The diameter, too, isn't critical, though I wouldn't go much below $1\frac{5}{8}$ in. if I were you; if you must, then I recommend that you widen the wheel to give sufficient mass in the rim, not forgetting that this may mean lengthening the crankshaft. The groove should be 45° included angle ideally, but your 55° Whitworth screwcutting tool should serve in default of aught else. There must be a sharp point, as it is important that the driving band doesn't bottom in the groove.

I have dealt with the method of flywheel manufacture without a lathe at the beginning of the book, but there is an alternative if you use a $\frac{5}{32}$ in. shaft, which is to use a Meccano wheel. Their No. 24 "Bush Wheel" with discs of brass about $2\frac{1}{4}$ in. dia. attached by six screws should do quite well, but you could use the No. 19A flywheel instead. This will mean using rather longer support pillars, shown in Fig. 2, and in such a case they ought to be rather stouter than shown, so that the support looks more in proportion. This wheel WILL just clear the steam pipes etc., and makes quite a companionable hum when running at speed!

Initial Assembly

For the two pillars Fig 3–2 I used a couple out of my bit-box – they came from some ancient dismantled wireless set. Cadmium plated, they polished up beautifully! However, there is a much easier way – use two pieces of 4 BA screwed rod of suitable length, and drill a

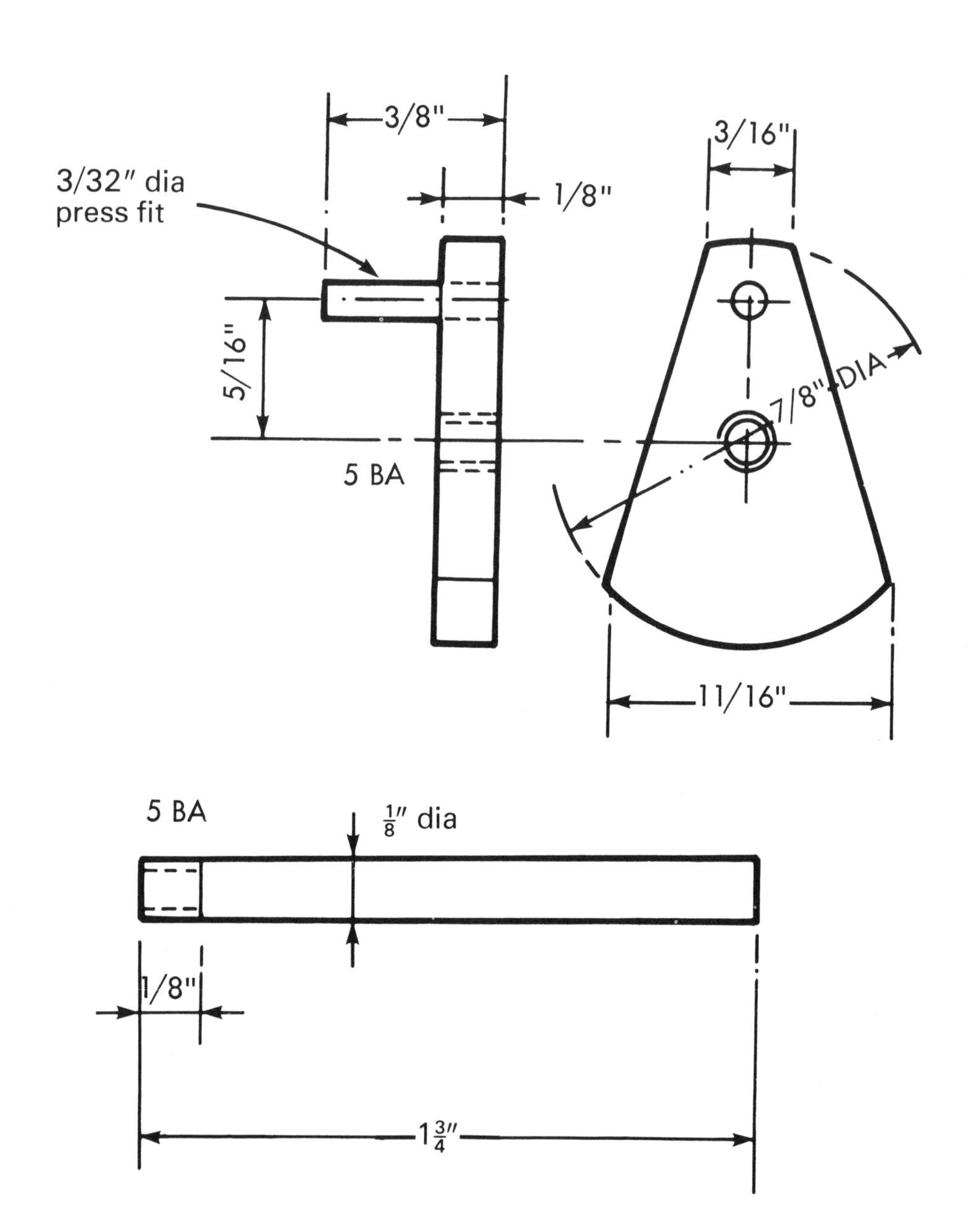

Crank Fig 3–3A

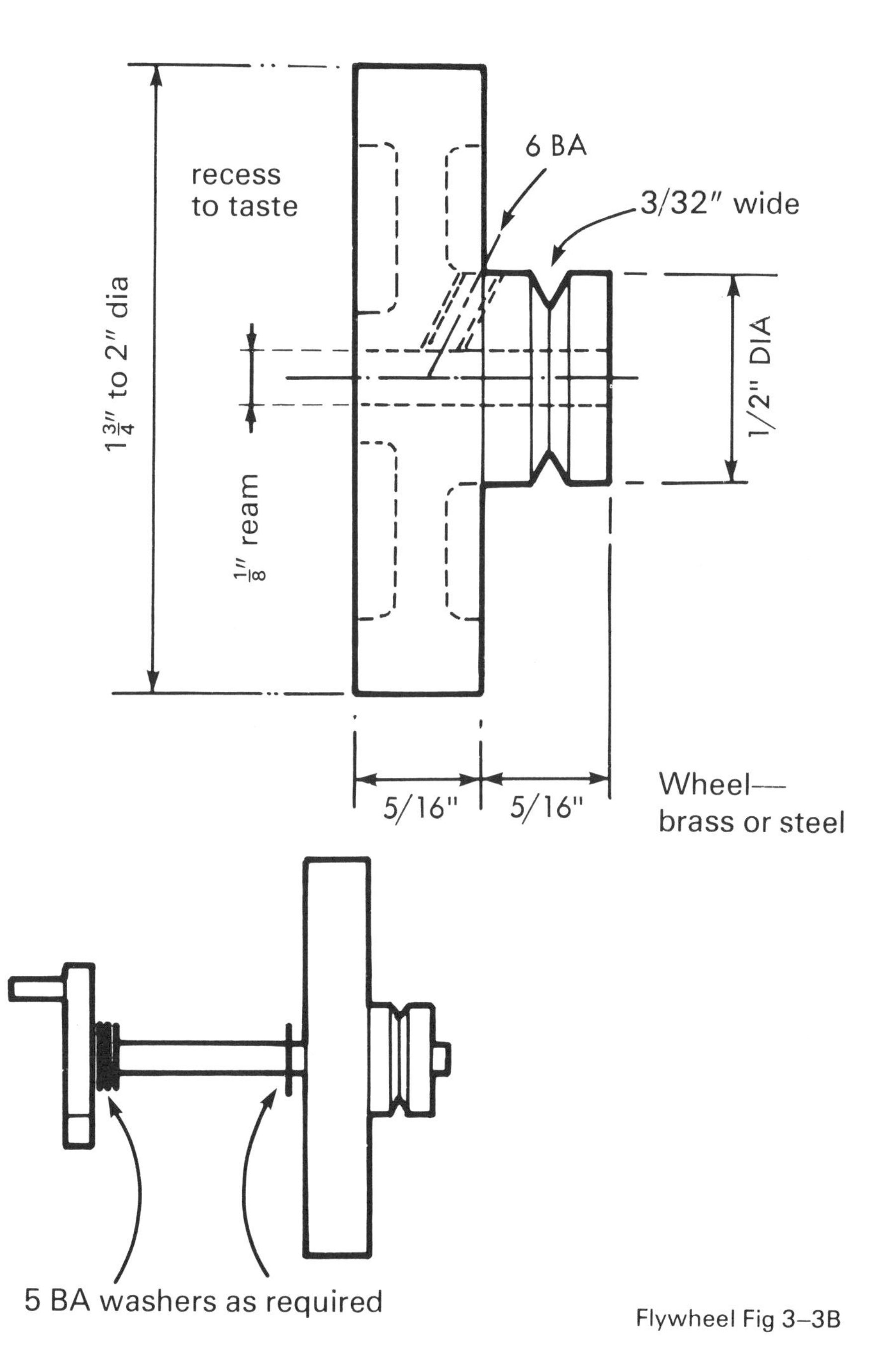

Flywheel Fig 3–3B

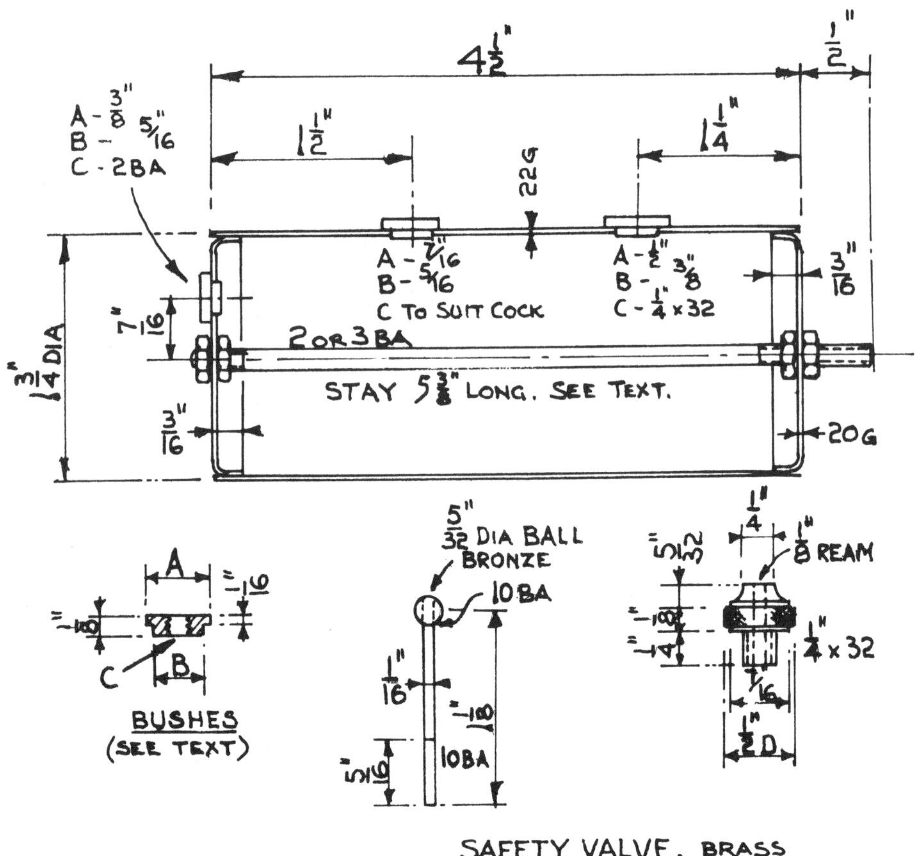

Boiler Shell and Details Fig 3-4

clearance hole right through two pieces of brass to form spacers. For that matter a piece of hard wood, say $2\frac{1}{4}$ in. long × $\frac{1}{2}$ in. wide and the right height would do just as well and be simpler! The steam pies are best fitted after making the boiler, but I had better mention them now. $\frac{1}{8}$ in. (3mm) steam pipe is quite adequate, but even this is too large to attach directly to the steam chest holes. I drilled about $\frac{1}{16}$ in. deep from the *back* of the angle-piece, $\frac{3}{32}$ in. dia. You must then file down the end of the steam and exhaust pipe to fit into this socket, and solder both in place. Alternatively, if you have any $\frac{3}{32}$ in. dia tube, solder in a couple of stub pipes, about in. long, to which the steam and exhaust pipes may be soldered later. Having fitted these, and mounted the frame on a temporary support (the vice, for example!) you can assemble. Fit the crank, and make sure it revolves freely with a trace of sideplay. Assemble the piston and cylinder with a few drops of cylinder oil, or SAE 30 motor oil, and see that this is free.

Fit the pivot stud to the cylinder with a drop of Loctite, and look for or make a little spring. You should be able to squeeze the spring flat between the finger and thumb of your left hand. (Or right finger and thumb if you are left-handed.) Attach the cylinder (conn-rod over the crankpin, of course), slip on the spring with a washer each end, then two lock-nuts. The assembly should rotate freely. If it doesn't, look for the fault and cure it. (Pivot pin not square is the usual trouble.) Rig up some form of adaptor to fit the steam pipe stub (the lower one) and attach a rubber tube to some source of air – tyre pump etc. The engine will need assistance to start, but should turn on about 5 lb sq. in. and run like blazes on 10 lb. Keep her running for some time and you should find that she will run much faster when she is bedded in.

Boiler (Fig 3–4)

This is quite safe for up to 60lb sq. in. or more, but pressed only to about 15. Nevertheless, I have included a stay for the flat ends in the design, to make doubly sure; it also helps in the assembly. Find your piece of tube, and a piece of copper, which should be 20 gauge or thicker, not thinner, for the ends. You will need some silver solder and flux – Easyflo No 2 is best – and this can be had from Messrs Reeves of Birmingham or Stuart Turners, Henley-on-Thames. Don't be scared of silver-soldering if you haven't done any before – it is, in fact, easier than soft soldering, and a little boiler like this can, with care, be 'brazed' up with the small 'Soudogaz' blowlamp.

File the ends of the tube square; it doesn't matter if it comes out $\frac{1}{4}$ in. longer, by the way – just takes a bit longer to get up steam, but stays longer *in* steam! Make a plug of hard wood (or metal) that is smaller than the bore of the tube by twice the thickness of the endplate. This is to beat out the flanges of the end-plates on. If of wood there is no need to radius the corners, but if you are turning it up out of steel, or have a piece the right size anyway, put a little radius on the corner. Mark out on the copper sheet two circles, one equal to the diameter of this plug, and another $\frac{3}{8}$ in. larger. Cut out two circles to this latter diameter, and then anneal them – raise to dull red heat with the blow lamp. Quenching in water isn't essential, but saves time. Hold the disc to the plug in the vice, truly central, and with a copper or plastic hammer beat round the rim to fold down the lip. Don't try to do too much at once – you will have to anneal the metal again about 3 or 4 times – but take each 'beating' several times *all round*, don't try to beat right down on one half and then the other, or you will get cockles. When you get nearly down, try the piece in the tube; it should be a push fit. At this stage you should beat out any cockles in the flat part, clean all up, trim the edges, and go all over with medium fine emery. Drill the holes for the stay and for the water-level plug.

Now drill or punch the holes for the bushes in the tube. If you have no lathe, you will have to buy these with the steam cock and safety valve; wait till they come to get the right size. The safety valve used on the Stuart 'S.T.' boiler will do very well, and their No 200/1 ($\frac{1}{8}$ in.) cock is the same as the one I used. The stay should be made of copper or drawn gunmetal, not steel or brass. The size is not critical – 4, 3, or 2 BA will do. Don't be tempted to use screwed rod – it will be strong enough but is liable to corrosion in the threads. Note that the thread is a bit longer one end than the other – this is to support the boiler in its casing. Run on a brass nut each end till the two are the same distance apart as the tube is long. Fit the endplates and put brass nuts outside. Insert in the tube, and make sure all fits closely; if there are any gaps, beat out the flanges a little. Silver solder won't fill gaps more than about .01 in. wide.

Take all apart, and clean everything thoroughly. Make a thinnish paste of flux and water and smear this over the flanges of the ends, about $\frac{1}{4}$ in. deep inside the tube ends, on the screw threads of the stay, on the shoulder of the bushes, and round the bush holes. Assemble whilst this is wet, making sure that the waterlevel plug bush is on the vertical centreline. (If you have no lathe, don't drill the hole for this bush. When you come to braze up the boiler, set a 2 BA brass nut in position, well fluxed, and braze this to the endplate. When all is finished drill through the nut 4mm and put a 2 BA tap through nut and plate together. This will be quite secure enough and just a touch with a file will give a nice flat seating for the screw plug.) Allow to dry and then wipe off any surplus flux that has run where it shouldn't; the alloy will follow the flux and you will get wasteful and unsightly smears of silver solder if you don't take care over this.

Make a little hearth out of old firebricks, preferably outside the back door, and get hold of some old asbestos pipelagging or string (NOT hard asbestos cement sheet; this won't stand heat). You won't need this if you have a 1-pint paraffin blowlamp, but will need some insulation to keep the heat in if using a small bottled gas blowlamp. The *very* small gas blowlamps won't do, I am afraid, but the Soudogaz is just big enough. Have a pair of tongs handy, and if you have abestos gloves or even an asbestos oven-cloth, these will be useful too. Set the boiler on end, front end up, and set asbestos sheet etc round it. Light the lamp, heat the end of the stick of alloy gently ($1\frac{1}{2}$mm or less is the right size, but it may be strip) and dip the end in the flux powder. Apply heat to the boiler end till it is dull red-hot – not bright red. Touch the joint with the end of the alloy rod, and this should melt and run into the joint. Follow round with the flame, touching with the rod to supply more alloy; the molten silver solder will follow the heat, and you should get a bright silver line showing right round. Apply the alloy rod to the bush and the brass nut on the stay, with a bit more heat if needed. Turn the whole the other way up and do the other end, but take some care to avoid filling all the threads projecting

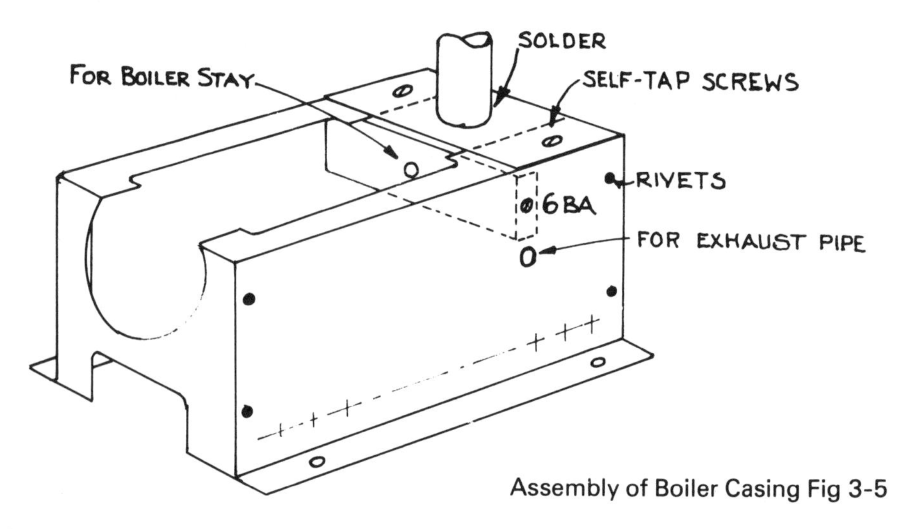

Assembly of Boiler Casing Fig 3-5

on the stay – just run a little where it meets the nut. Turn on its side and braze in the two top bushes.

The most important part about this operation is that you must keep the end of the rod well fluxed. Keep dipping it into the powdered flux as you go along, whilst the rod is hot. Furthermore, if you notice that the alloy isn't running well, use the rod to apply a little more flux at that point. Extra flux does no harm, though if you use so much that it runs down the work the brazing alloy may follow it. As I have said before, if you feel at all nervous about it, it undoubtedly pays to practise on a few bits of scrap copper first.

Let the job cool till it is 'black hot' – fizzes when you spit on it – and dump it into a bucket of water, taking care not to get a spit in your eye! This will clean off the scale and if you leave it there for 20 minutes or so the flux residue also. Pickling is better, but not essential, Take the boiler out, and if there is any flux left on, scratch it off with a wire brush. Examine the job all over, and if there are any bits that have missed the solder, reflux the whole of that joint, repeat the process, concentrating the heat on the missed part and applying more alloy. Finally, clean off any surplus alloy and after cleaning out the threads in the bushes, polish all.

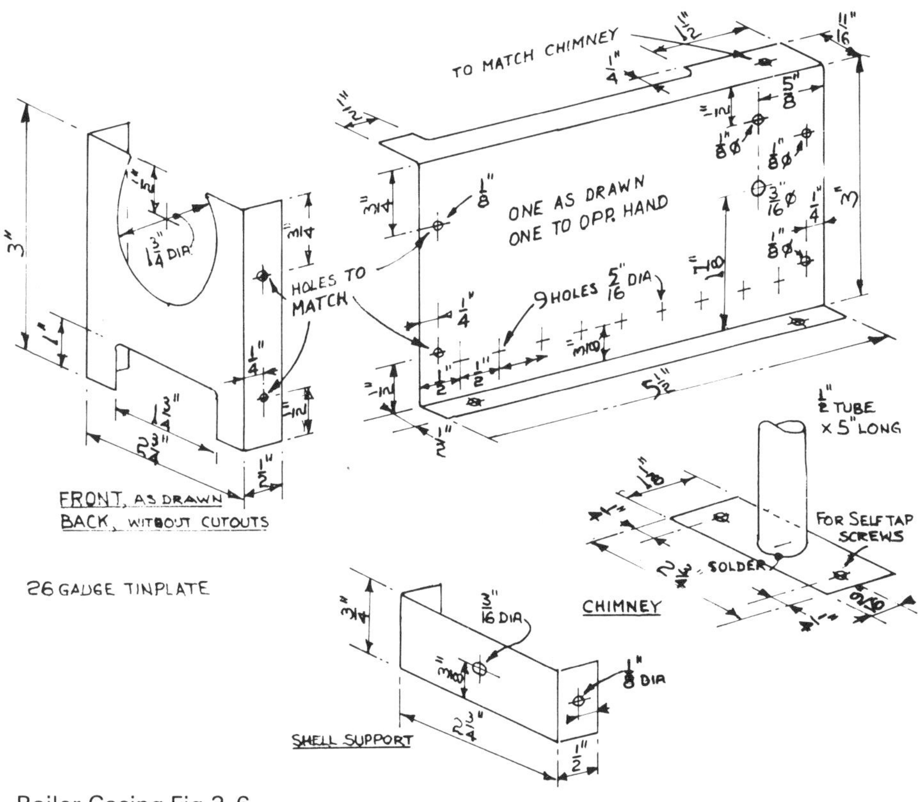

Boiler Casing Fig 3-6

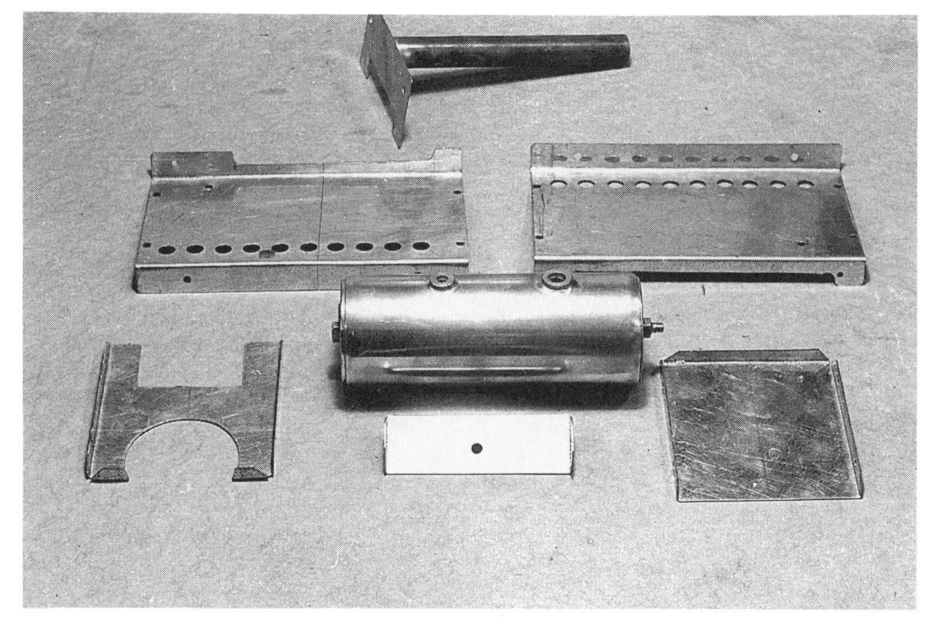

The finished boiler shell and the parts for the casing.

Boiler Casing

The boiler case (Figs 3–5 & 3–6) is folded up out of tinplate – 26 gauge is quite good enough; it stiffens up remarkably when folded and jointed. You can join the parts with nuts and bolts if you like, but I used pop-rivets, apart from the plate that carries the chimney where I used self-tapping screws, and the cross-bar that carries the boiler end, which has 6 BA nuts and bolts. Take care that you get the folds square. They can all be done in the usual way with smooth pieces of steel angle set in the vice; no problem. Don't fit the boiler shell till you have tested it. The photo shows the parts for the casing.

Safety Valve

To make the safety valve (Fig 3–4) chuck a piece of $\frac{1}{2}$ in. round stock, turn the end down to $\frac{1}{4}$ in. about $\frac{1}{4}$ in. long, and screw to $\frac{1}{4}$ in. x 32 tpi. Turn the curved profile – I used the back of a curved facing tool for this – and then drill 3mm. Knurl the $\frac{1}{8}$ in. wide shoulder, and then part off. Ream the hole $\frac{1}{8}$ in. and lightly countersink the lower end of the hole – leave the top a sharp corner. Grip a $\frac{5}{32}$ in. ball (bronze, not steel) in the chuck, just touch with a centre-drill, and then drill and tap 10 BA. It doesn't matter if you go right through, but I went in $\frac{1}{8}$ in. only. Make the stem of the valve from a piece of $\frac{1}{16}$ in. brass rod, and screw one end into the ball. You need a washer with a *square* hole in it, (a) to stop the spring from going through the $\frac{1}{8}$ in. hole and (b) to let the steam through. I filed a square in a 10 BA washer. To set the valve, make a spring from say 26 gauge bronze wire wound round a $\frac{3}{32}$ in. rod, about 10 coils or so. Assemble the valve, and adjust the spring till, using the kitchen scales, it takes between 2 and 3 onces to lift the ball from its seat. This will give between 10 and 15 lb sq. in. Then, lightly tap the ball with a hammer to form the seat. Note that the washer with square hole goes on first, then the spring, then a plain washer, and finally the nuts.

If you have no lathe then, of course, you will have to use a commercial safety valve and, as noted earlier, will have used the bush supplied with it when making the boiler. The main problem with such valves is that they are usually set to a higher pressure than we need. Ask the makers to declare this pressure and, if more than 20lb. sq. in. you may have to make a new spring. It is no use fiddling with that in the valve – nothing you can do to it will alter the set pressure. A new spring of thinner wire is the only solution. The suppliers may, of course, be able to supply a valve for this pressure if enough of you decide to make the engine so that it is worth their while!

Boiler Testing

To test the boiler you really need a little pump and a reliable (not model) pressure gauge. You may not have one of these, but it is worth enquiring in the local M.E. Club if there is one, as most clubs nowadays have this equipment, in order to comply with the requirements for running model locomotives. Failing this, you will have to improvise. If you live in a town with a good water pressure (a phone call to the local Water Authority will tell you this) then you can manage very well. The boiler will stand 60lb. sq. in. by design, but works at only 15, so that a test at 30lb. sq. in. is sufficient. That is about 70ft. head of water – say 20 metres. So long as the mains pressure doesn't exceed 60lb or 140ft head, you can safely connect to the tap. (Note, though, that if you are in a block of flats you are probably supplied from a roof tank, and may have to go to the basement to do the test!)

Fit a plug to the level gauge hole and screw in the steam cock. Get a short length of tube that can be screwed the same size as the safety valve and fit this to the bush. Use a bit of jointing on the threads – paint that has gone a bit thick, or putty let down with a bit of white spirit will do. Fill the boiler, letting air out through the steam valve, which then close. (You may have to wangle a bit of rubber tube from somewhere to fit both the tap and the screwed pipe you have fitted) Dry the boiler with a towel and look for leaks. If it spurts out all ways, then you will have to start again! But if there seems to be only a "bit of a weep" dry with the towel again and mark any definite leaks. (If there are NO leaks, leave it there for 10 minutes to make quite sure, and then raise a cheer!) Mark any pinholes you find with a felt pen or a pencil.

If it is only a tiny pinhole, then you may be able to cure it with a small dab of tinman's solder, but if more than that I would prefer to rebraze that part. Flux ALL the joints at the affected end (let the water out first, of course) get the job hot again, but this time concentrate the heat only at the defective part of the joint. Apply the fluxed rod – the alloy will run a little to either side and you will then know that you have a sound joint. Clean off as before, and repeat the test. You may think all this is making a big fuss about a little leak, but "model" steam is just as hot as the real stuff, and when one is making something that will be handled by children it is just not permissible to take risks.

You can now assemble the boiler to its casing. The chimney is an "optional extra" – mine is made from a bit of scrap $\frac{1}{2}$ in. water pipe, and since the photo was taken I have bell-mouthed the top a bit. It is soldered over a hole in the top-plate; quite adequate, as it doesn't get very hot. The exhaust steam pipe sticks up inside the chimney by about half an inch.

Burner (Fig 3–7)

This is a matter of choice. I don't like meths with small children – all too easy to

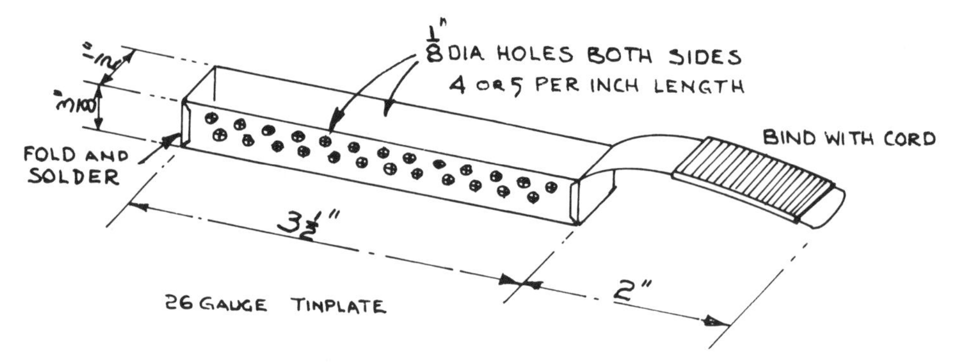

Burner Fig 3-7

spill some and have it set alight. The fact that the flame is almost invisible adds to the danger. On the other hand, 'META' fuel is dangerous if eaten. My choice is for META, as my own offspring is well disciplined over 'things that look like sweets', but even an experienced adult can tip over a bottle of meths. The burner is very simple – just a trough with air holes, 5 – $\frac{1}{8}$ in. holes per inch length each side. The META bricks are broken in thirds, and alternatively stood on end and laid flat edgeways in the trough. This provides an adequate flame – if too adequate, lay the alternate ones flat. There is enough water in my boiler to stand one recharging of fuel, and still leave it $\frac{1}{4}$ full.

Final fitting up

My original base was a piece of wood, with asbestos millboard to stand the fireholder on. But this will, in due course, be replaced by a sheet of steel plate, about 24 gauge, with the edges punched with $\frac{5}{32}$ in holes at $\frac{1}{2}$ in. centres to match the Meccano which my daughter uses. A few such holes in the vicinity of the engine shaft could enable the engine to be geared to a Meccano drive if desired.

Piping up is a matter for experiment, once the points I mentioned on page 56 have been dealt with. The best thing to do is to set up the engine and boiler on its base and then bend up a piece of soft wire to get an idea of the run first of the steam and then of the exhaust pipe. With these as templates you can bend copper pipe ($\frac{1}{8}$ in. or 3 mm. for steam, $\frac{1}{8}$ in. or $\frac{5}{32}$ in. for exhaust) leaving this a trifle longer than needed. Offer these up till you get a smooth and sightly run to the pipes. The tube may need annealing; heat to red and, in this case, quench in water to remove the scale and make it easier to polish. Once you have the pipe run right, clean it up; first solder on the union for the steam cock, then the other end to the port block (I have mentioned this before) The *lower* hole is the steam inlet, by the way. The exhaust pipe is done in the same fashion, except that it just pokes up into the chimney (if fitted) for half an inch or so, passing inside the boiler casing to get there. You could leave it off altogether if you liked.

A little tray under the cylinder might be a useful addition, to catch the oil and water drips, and make peace with the domestic authorities. One point – why not fit a *cock* instead of a 2 BA plug for setting the water-level when filling? (Don't forget to make a little funnel). Well – It isn't really safe to fit a cock for setting the water level. Your own child may know not to turn it when in steam, but young visitors can't resist turning on or off everything in sight, and this could cause a nasty scald. A screwed plug is less temptation, and less easy to undo, too; it's hot!

I gave the boiler casing two coats of aerosol red primer, and one coat of red paint on top, but black for the chimney. This has, so far, stood up to the heat very well. As to time, it shouldn't take you more than three evenings, or perhaps four, to complete – less if your scrapbox is more rewarding than mine! Actual construction took very little longer than looking for bits; if you made it of all 'new' material you shouldn't take nearly as long.

You may find the engine a bit stiff to start with and it pays to use an excess of oil at this stage. However, don't put so much down the cylinder that it gets full up or you *will* have trouble! It pays to use water out of the kettle, as most tap water contains some degree of hardness, as well as air, and both can be a nuisance. Hot water also saves a certain amount of fuel. If you find there is a persistent leak of steam (or water, which is condensed ditto) at the portface check (a) that the pivot screw is square to the cylinder face – if it is not then it will hold the faces apart. (b) that the crankpin is not bent in relation to the shaft; if it is, you will almost certainly see the big end wobbling from side to side. If pronounced, this wobble can pull the cylinder faces apart. In each case *careful* bending may correct the fault, and in the second situation a slight enlargement of the big end hole may help. (Don't go too far, or she will "knock"!)

If there seems to be a lot of steam passing the piston the best thing is to make another – it's not a big job – and make it a closer fit. The snag here is that there is some differential expansion when hot, and (especially with stainless steel) the piston clearance enlarges. Further, unless you really polished the inner surface of the cylinder the manufacturing roughness wears off fairly rapidly and again increases the clearance. The ultimate solution is to use the packed piston; the groove in which should be filled with cotton thread well soaked in tallow or engine oil. The best thread to use is obtained by teasing out the core of the small round lamp wick used in the paraffin nightlights. Or, of course, the "pukka" graphited yarn if you have any. Don't make the packing too tight, but there should be sufficient put on to cause a slight resistance but with no metal-to-metal contact. However, don't get worried about such leaks till the engine has been run about half-a-dozen times, as many may be cured by bedding down.

Finally, if you are making this little engine as a present for a child, make out a proper set of "Engine-driver's Rules" – as were to be found in all engine houses – telling him or her exactly what to do, not forgetting a warning about the META fuel *and* about letting all cool down before opening up to put more water in. Find a little bottle (those used as "Miniature" Drambuie suit fine) for oil; you can get special cylinder oil in small tins from Stuart Turner Ltd, but ordinary SAE 30 (or W20/40) will do. Find a screwdriver for the level plug and, above all, don't forget that the shops are shut on Christmas day, and they will need a packet of META, too! A final polish, a nice box, and there you are; a present no child will ever forget. And perhaps one for yourself, as well?

Chapter Four
HERCULES
A working model steam crane.

Whilst the previous engines have been built from bar stock, this model uses castings for the engine proper, obtainable from Messrs A. J. Reeves & Co., Ltd., Holly Lane, Marston Green, Birmingham and marketed as their "Popular" engine. In ordering you should mention that you require these castings for this Crane, when they will package up the parts required – though the design uses two cylinder sets you will need only one flywheel, for example. The gears are "MECCANO", obtainable in large towns from toyshops, and also from specialist MECCANO suppliers.** However, there is no reason why you should not use other gears if you have them available, or even some from an old clock; the latter will, however, be a bit narrow and you will have to be content with a smaller load capacity. It is only necessary to set out the gears on a piece of card and adjust the centres, then altering the drawing to suit – but bear in mind what I said on page 10 about altering dimensions!

The jib is specified only in outline on my drawing. My own is, in fact, made from a couple of brake rod fork ends brazed to a length of steel tube. However, you can use brass curtain rail of "I" section separated by stiffeners, or brass angle with latticework, or even hard wood; not an unusual jib material on smaller cranes, usually octagonal in section. So long as it is about a foot between pivot and pulley, and looks in proportion, you can make it to suit your own taste.

The "engine" can be a twin-cylinder, double-acting, cranks at 90°; twin single-acting, cranks at 180°; or even a single cylinder, double-acting on one side only. Mine has the twin DA set-up, and doesn't really need a flywheel, but the others will; so I put one on mine to make sure it would go in. The type of engine used won't affect the lifting capacity all that much, as the limit here is what it will carry without toppling over, not engine power. Mine will lift over 10 lb. with 10 p.s.i. of steam – if I clamp it down.

The boiler is a very simple pot, with a steam drying coil. The main problem is not *making* steam, as much as reducing blow-off; steam demand is intermittent, and there can be quite lengthy idle periods whilst loads are hooked on and balanced (a crane teaches a youngster quite a lot about centres of gravity and such) so I have called for a "top-hat" damper to go over one of the wicks to reduce fire in such periods. But you might care to consider using Meta fuel, and give the youngster a bit of experience in managing a fire. I *don't* advise this if there are any toddlers in the

**

The MECCANO works in Liverpool is now closed but the scheduled gears are still available in the larger shops. In case of difficulty write to: M W Models Ltd., 4 Greys Road, Henley on Thames RG9 1RY *or* to AIRFIX Ltd, Haldane Place, Garrett Lane, London SW18.

Hercules

family as the tablets look too much like sweeties and won't do the tummy any good. You will notice I have called for a screw-plug as a water-level indicator when filling the boiler. Most people put a drain tap here, but I prefer a plug as there is less risk of the tap being opened when under steam; even at 5 p.s.i. this can give a nasty scald.

None of the unmarked dimensions is critical and you can adjust many others to suit yourself or the contents of your workshop. The base wants to be either large or fairly heavy (one I made 25 years ago was mounted on an old cast-iron pulley) and the pivot for rotating the crane should be more or less underneath the winding drum – forward of centre of the baseplate, that is – to give a bit of counterbalance. The rotating base can be of wood if you like, but I used a bit of steel plate well painted against rust from leaks. The "works" are thus located with "proper" bolts which won't work loose over the years as woodscrews might. In which connection, try to use nuts and bolts which will be easy to come by; e.g. 2-4-6 BA rather than 3-5-7; or even Meccano ones where suitable. Use screwdriver slot bolts – this is a case where they are entirely appropriate, as it makes running repairs and periodic overhauls easier for the lad (or lass!) and it's better that he gets this sort of experience than that he has true-scale hexagons all over. So, to work.

Engine standards Fig. 4–1

Reeves supply part-sets of the engines. For the two-cylinder job you require parts 3, 4, 5, 6 and 9; and two stands, part 10. If you decide to fit a flywheel (you must for the single-acting designs) this is part 1. I also suggest you obtain the screws, part 14. The drawing supplied by Reeves is part

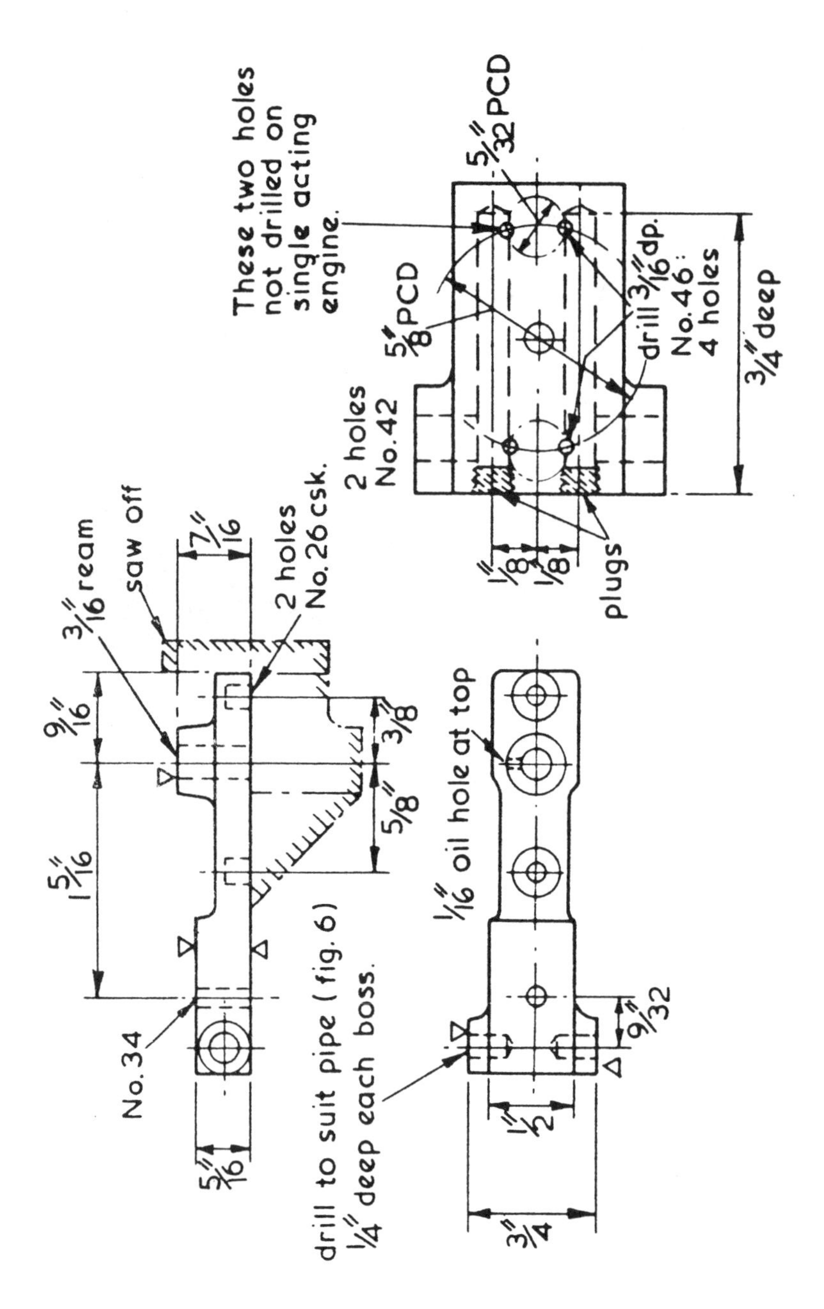

STANDARD: 2 off opposite hand Fig 4–1

RV23, and it will help to have this, if only to save mucking up the book in the workshop. However, it differs from mine so start by making the following alterations. *Stand.* Steam inlet and exhaust bosses to be $\frac{5}{32}$ in. drill (or to suit your steam pipe), not $\frac{3}{16}$ in. x 40; steam and exhaust ports drill No. 46, not No. 50; main bearing hole, $\frac{3}{16}$ in. *Cylinders.* Drill the ports No. 51, not No. 52. These changes will improve the performance of the engine on load. Please note all these changes are incorporated in my drawings.

Saw off the foot of the standard, leaving as much of the column available as possible, and also the bearing boss and rib at the back. File or mill the back dead flat and trim up the sawn-off end. Now mill or file the portface parallel to the back face – keep reasonably near to the $\frac{5}{16}$ in. dimension. File the boss of the main bearing to $\frac{1}{8}$ in above this portface. Mark out the longitudinal centreline and fix the position of the crank bearing. From this, mark out for the pivot-pin hole (No. 34) and from this scribe at $\frac{5}{16}$ in. radius to find the centre of the ports. Mark these out very carefully start with a *very* fine dot-punch, enlarge the dot with a tiny drill, and then very carefully drill them No. 46. The idea of drilling the ports first is that if you make a boss-shot and have to plug and re-drill you still have the pivot pin centre to mark out from again. Drill the pivot hole, the two No. 23 fixing bolt-holes, and drill and ream the bearing. Put in a $\frac{1}{16}$ in. oil hole. (Get it right way up! The two frames are to opposite hands).

Then mark out for the No. 42 passages, which are not required on the single-acting engines, and drill these, finally drilling the bosses so that they fit whatever size steam pipe you decide to use ($\frac{5}{32}$ in. is all right) through to meet the passages. Then tap the no. 42 holes 6 BA just a few threads, screw in a short length of 6 BA brass, and seal with soft solder. Blow out all holes well with compressed air or a cycle pump.

Cylinder Fig. 4–2

Clean up the casting all round with a file, and when doing so try to get the end as square to the body as you can – saves a bit of trouble later. Chuck lightly in the 4-jaw and estimate how much must come off the flanges to get both ends equal. Face the flange to this figure. Centre, and drill $\frac{5}{16}$ in. right through; take care that the drill doesn't snatch – I have a set of straight - flute drills for brass and cast G.M. Set up a boring tool and rough bore to within a few thou. of dimension. Then rehone the tool point and take light cuts on fine (power) feed till you reach dead size, which you can determine by measuring the shank of a $\frac{3}{8}$ in. drill – or $9\frac{1}{2}$ mm if you work metric! Don't be scared of power feed in a bore – if you are doubtful, use the tumbler reverse to traverse away from the chuck, and set the cut on with the tool right in. If you want to ream, then leave the bore 3 thou. small and then hold a reamer in the tailstock chuck and pass through by sliding the tailstock bodily, running at about 100 r.p.m. Whilst in the lathe, set the portface vertical and scribe a centreline as far as you can round the casting. Put a centrepop on the outer end, to show that this face was machined at the same setting as the bore. This end will have the gland cover.

Reverse in the chuck, and with a bit of $\frac{3}{8}$ in. stock in the bore, set to run truly. Face this end also.

Now, the portface must be parallel to the bore. If you have a vertical slide for the lathe, and a milling cutter or flycutter, you can set the casting on a mandrel and grip it in a vice on the vertical slide, using the mandrel to set the bore square across the lathe bed, and then face to dimension. If you haven't one, then set the mandrel between centres and use your scribing block to mark out round the casting at $\frac{7}{16}$ in. from the bore centre. File to this mark. To ensure parallelism, set the casting face down on the lathe bed and use your scribing block to ascertain whether the face is parallel, adjusting your filing to suit. Finish by scraping the face flat using a piece of mirror glass or the lathe bed with some marking blue. (Prussian blue oil paint from a tube will serve if you have none.)

Return to the lathe, and set the mandrel between centres. With the scribing block set at centre height and the portface set vertical with your square, lightly scribe the centreline. Now set the cylinder vertically on the gland-cover face, and mark out for the three holes. Actually, it is better to mark out for the pivot hole only, and for the ports by using dividers set to $\frac{5}{16}$ in. and strike a radius from the pivot centre, but not everyone has toolmakers dividers. (Worth getting a pair, in that case. They are an essential tool.)

Drill for the pivot hole taking care (a) to drill as square to the face as possible – this is important; and (b) not to go too deep. Tap the hole equally carefully, and don't force the tap against the bottom of the hole, or you will make a bulge in the bore. Drill the two No. 51 holes, and then file a tiny channel in the bore from the emergence of the hole to the end of the bore. Finally, file the relief on the surface, about 10 thou. deep, leaving a face $\frac{3}{16}$ in. wide in way of the ports. This ensures that the cylinder beds firmly against the port faces in operation. Remove all burrs, and check that there is no bulge in the bore, mentioned above; if there is, then you must either file it away with a half-round Swiss file, No. 4 cut, or use a reamer.

The cylinder covers should present no problem, though the back cover is a pig to hold in the chuck. If you face one side – the outside – you can then solder it to a piece of brass, hold this brass in the 4-jaw, adjust to truth, and then machine the inner spigot and the O.D. The spigot should be $\frac{1}{16}$ in. proud.

You can then hold by this spigot in the 3-jaw chuck to machine the little recess. At the same time, mark out for the 3 holes; simply set each jaw of the chuck horizontal in turn with a spirit level and scribe across at centre height; and then mark a radius on each line at $\frac{19}{64}$ in. ($7\frac{1}{2}$ mm) below centre.

The Reeves drawing shows a fiddly little gland, but I didn't use one. You may do either, though at the low pressure the engine uses, a ground rod working in a reamed hole seals well enough. Machine the O.D. of the gland and the outer face first. Then reverse and hold the piece by the gland boss to form the spigot and drill the hole; drill No. 31 and then ream $\frac{1}{8}$ in., but if you have no reamer that size enlarge with a $\frac{1}{8}$ in. drill, running very fast and feeding very slowly. Machine the O.D. of the flange to size and then mark out for the screw holes as before. Drill the holes in the flanges. A tip; if you hold these on the machine with your paws whilst drilling, the drill will take charge and you will lose part of a finger. But how to hold them in a vice? The trick is to drill a $\frac{3}{8}$ in. hole in a piece of wood, square to the face, slit the wood block down the centre and set this in your vice; then grip the cover spigots in this $\frac{3}{8}$ in. hole whilst drilling. This done, offer the covers to the cylinders, setting one hole diametrically opposite the ports and spot through. Mark each cover to its own cylinder end, and then drill and tap the cylinders 8 BA.

Piston and Rod Fig. 4–2

I suggest you make the piston of brass. It is not a difficult machining operation; chuck a piece of $\frac{7}{16}$ in. or $\frac{1}{2}$ in. dia. stock in the three-jaw chuck and rough down about $\frac{1}{2}$ in. of length to 10 thou. oversize. Now fine machine until the first $\frac{3}{16}$ in. will just enter the larger of the two bores – if they are different. Take off another thou. or thou-and-a-half. Centre the face and drill No. 37, then tap 5 BA guiding the tail of the tap in your tailstock chuck, the tapwrench being set on the plain shank. (I have ground a little flat on all my "piston size" taps for better grip when using this method) With a parting tool form the groove; remove the burrs from the edge, and then part off the piece $\frac{5}{32}$ in. thick. Repeat for the second piston, and don't forget to mark each to its appropriate cylinder.

The shank of the rod is no problem; cut off to correct length, and then screw each end using the tailstock die-holder. You must run a file over the tips of the threads at the "big end" as the extrusion which occurs when tapping may cause the threads to be oversize and they won't go through the gland. For the big end bearing I suggest phosphor-bronze if you have any, or drawn gunmetal, but brass will serve, though won't have as long a life. (Perhaps 20 years instead of 50!) Make it from quarter-inch stock, polishing the surface first, and after turning down the little boss, drill and tap 5 BA as you did for the piston. But try not to go more than $\frac{7}{32}$ in. deep, to avoid fouling the cross hole. Remove from the lathe and file the flats as shown. Then make a trial assembly of the piston, rod, and big end, and mark out for the hole at $1\frac{5}{16}$ in. from the piston face. Drill this No. 31, and follow with $\frac{1}{8}$ in. to bring to size – I know the drawing says "ream" but this isn't really essential. In fact, if you drill $\frac{1}{8}$ in. at one go, this will probably give you the right clearance, as most small drills drill large. Lightly countersink the ends of the hole.

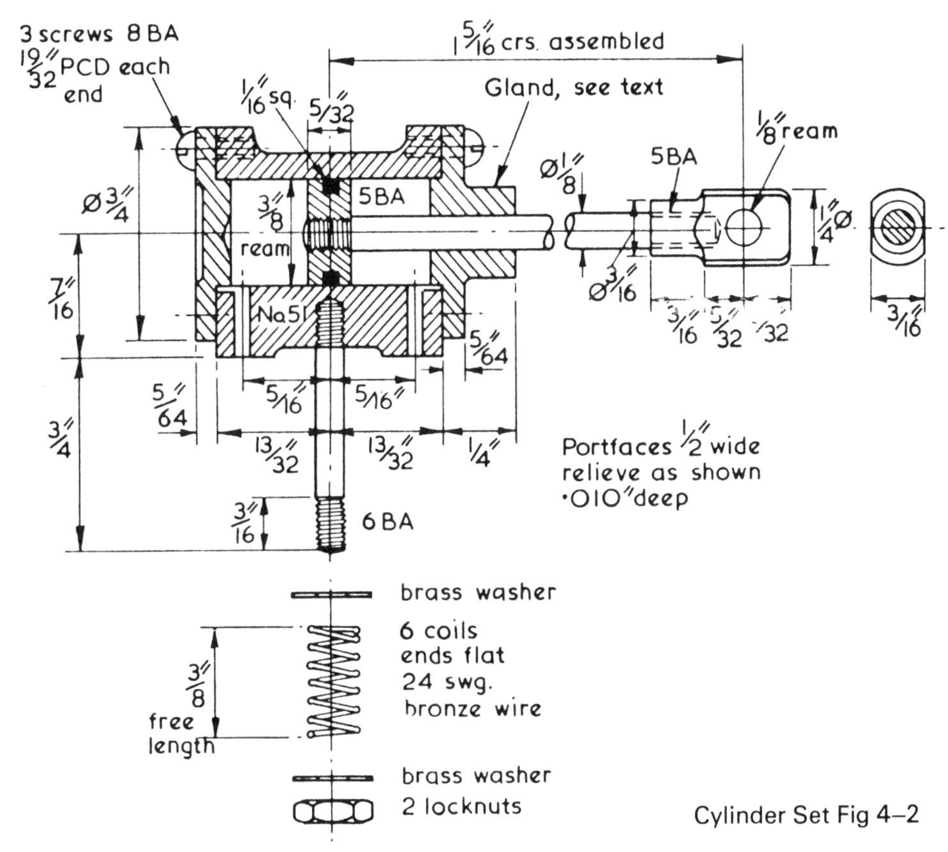

Cylinder Set Fig 4–2

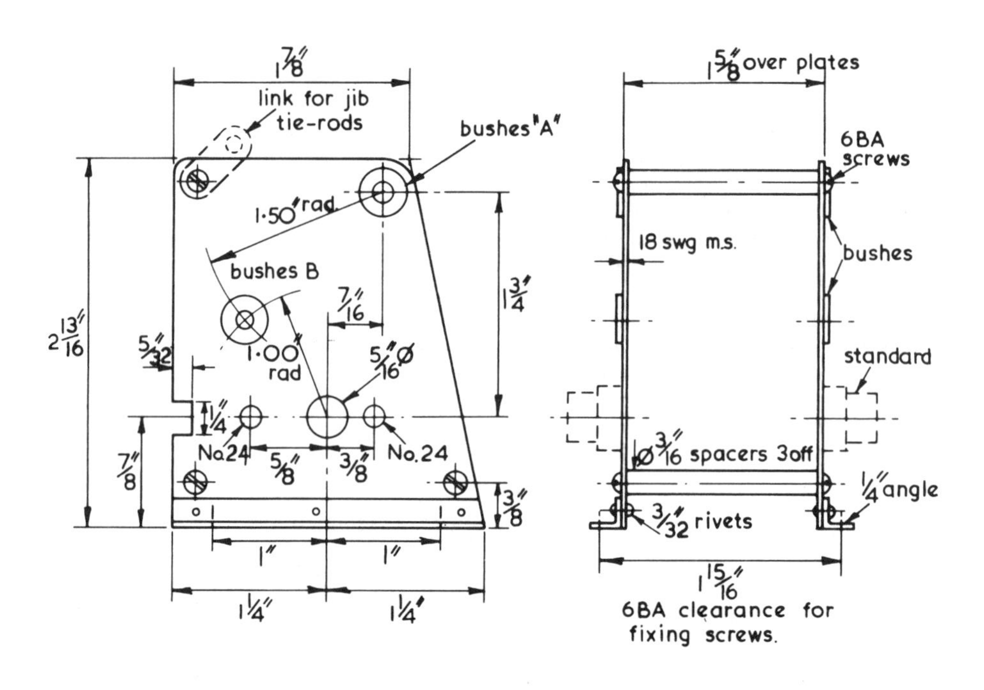

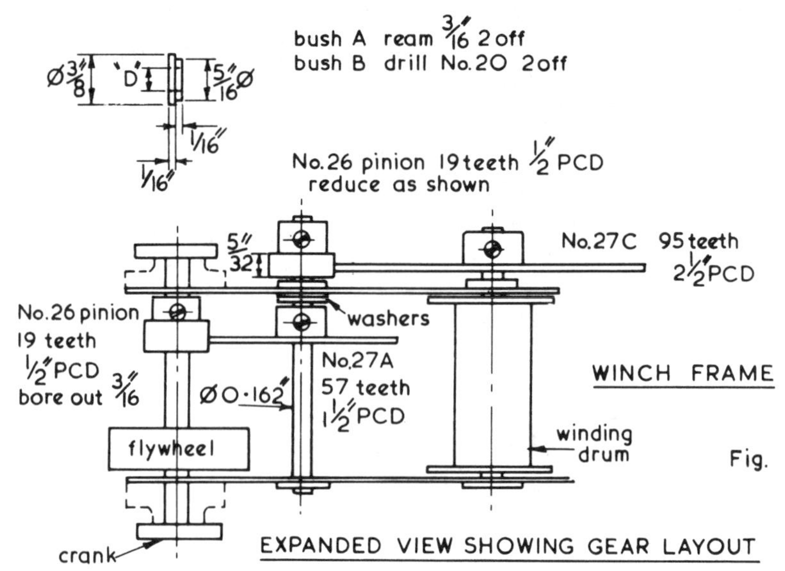

Winch Frame Fig 4-3

You won't need reminding that it is important that the hole be square to the axis of the rod. To ensure that this is so, when holding in the drill vice, screw in the shaft and use a square to check that it lies at right angles to the drill itself.

Winch Frame Fig. 4–3

Either brass or steel will do – I used 18 s.w.g. steel for economy! You may feel that if you use brass the bushes won't be necessary; you could be right, but don't forget the crane may well pass on to your great-grandchildren! Cut out the frames to the shape shown – or to suit your own ideas – and I recommend the use of a saw rather than shears. Square up the bottom and back edges to act as reference faces when marking out for the holes. Locate the centre for the crank first, then for the winding drum, and from these two for the second-motion shaft. The sizes and Meccano Cat. No. of the gears are shown on the drawing – you need two pinions nominal $\frac{1}{2}$ in. dia., one gear nominal $1\frac{1}{2}$ in. dia. and one nominal $2\frac{1}{2}$ in. dia., which give an overall gear ratio of 15:1.

Mark out for the other holes, and drill those marked 6 BA clear. Bolt the two plates together and drill the other holes, enlarging the bush-holes step by step till the right size is reached. File out the $\frac{1}{4}$ in. x $\frac{5}{32}$ in. notch at the back – this is to clear the spring of the cylinder pivot pins. Make the bushes to fit the holes, the width (thickness) being such that the bush projects a few thou. through the sideplates. Remove all sharp edges from the plates, more than mere de-burring, to avoid risk of the child cutting itself. Fit the bushes with a touch of Loctite and leave to cure.

The spacers are made from round or hexagon stock (polish them up a bit) parted off to length – take care to get them all the same – and the ends drilled and tapped 6 BA. Lightly countersink after tapping. Assemble the whole and check that (a) the frame sits on a flat surface without rock and (b) that a spindle passed through the bushes rotates fairly freely. If you have trouble, correct the fault. Now attach the two standards using 4 BA screws and make sure that a $\frac{3}{16}$ in. spindle will rotate in these. You *may* have to draw the fixing holes a trifle to achieve this, but if it binds no matter what you do, then it's probable the plates are not parallel to each other and you'll have to put that right by adjusting the spacer lengths. Once you are satisfied, measure the distance over the crank-bearing bosses and note this figure for making the crank. Remove the standards and make and fit the angle pieces by which the frame is fixed to the base.

Crank and Spindles Fig. 4–4

The drum spindle is simply a piece of $\frac{3}{16}$ in. silver steel with the end turned down a good fit to the largest gearwheel. (The tolerance on Meccano gear bores is $\pm$0.001 in. on 0.162 in. dia.) The intermediate or second-motion shaft can be a plain Meccano axle (these are made from nominal No. 8 s.w.g. coated wire) but I find that the coating doesn't like running in a well-fitting GM bush and soon wears slack. So I turned this shaft to 0.162 in. dia. from $\frac{3}{16}$ in. steel and then polished it so that it was both a good fit to the gearwheel and a nice running fit to the bushes.

The crankshaft is turned so that the distance between the shoulders is that shown, *provided* that this is at least $\frac{1}{64}$ in. more than the distance over the bushes mentioned earlier. Make it the greater of these two figures. Part off to length, then turn down each end to $\frac{1}{8}$ in. dia. x $\frac{3}{16}$ in. long. The ends must be screwed with the tailstock dieholder, as close to the shoulder as you can. The disc is faced in the lathe, drilled and tapped from the tailstock, and then marked off using your height gauge for the crankpin before parting-off. Turn round in the chuck, and run a No. 30 drill in for a couple of threads. Drill for the crankpin, taking great care that the hole is

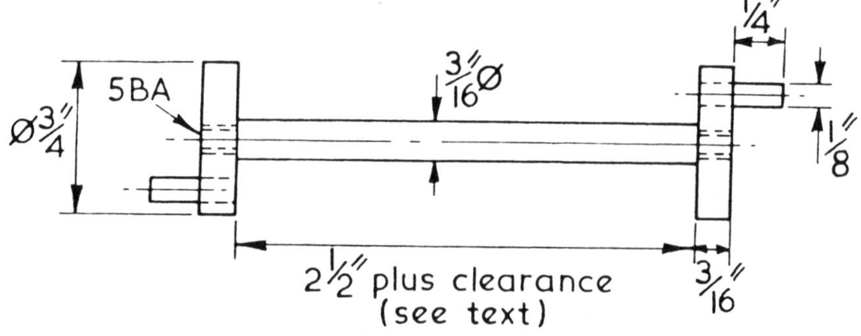

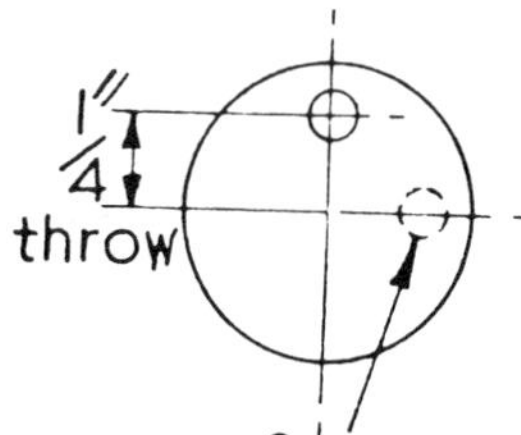

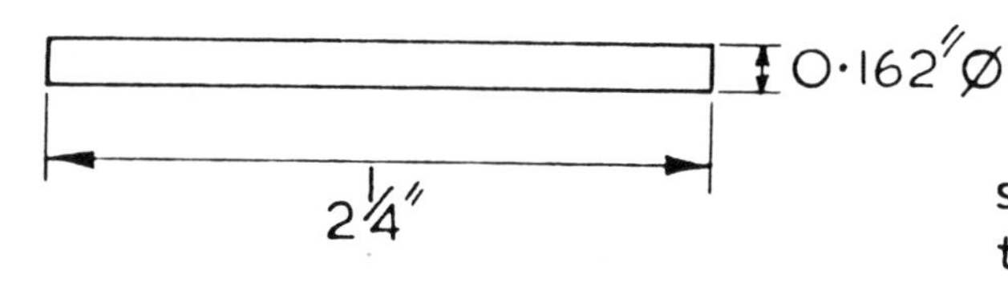

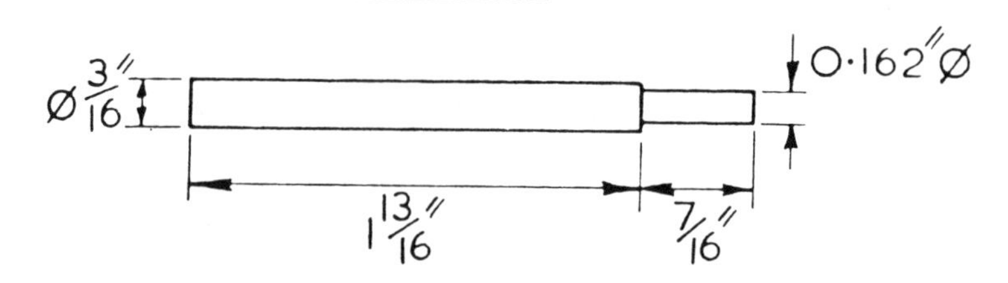

Crank and Spindles Fig 4-4

truly vertical to the face. (Chaps often sneer at oscillating cylinder engines, forgetting that alignment is ten times as important for them as it is for slide-valve engines, where you can get away with almost anything! You need *good* workmanship to make a good oscillator.) Make and press in the crankpin, using Loctite if need be. Now offer up the crankdiscs to the shaft. If the discs don't come at the right angle, carefully file the back face of the disc till they do – 90° or 180° as the case may be. If it's only a few degrees away that doesn't matter much; if a trifle more you may be able to force it; but if you need to go more than 15° or so, file it. Note, by the way, it doesn't matter which crankpin leads; though in real practice the right-hand engine was usually the leader.

Now, of course, you are certain you have (or I have) made a mistake, because the $\frac{1}{2}$ in. gear won't go on the shaft. I know; you have to bore it out to fit! Grip lightly in the 4-jaw by the boss and adjust till the bore runs true, using a bit of axle to help you set it. Then carefully enlarge the bore with successive drills and finish with a $\frac{3}{16}$ in. reamer. Take the grub-screw out first, though! You can now offer all up and try the gears for mesh. If there is slight binding here or there, ease the top of the teeth of the offending gears *very slightly*. If they are miles out, then you have the second-motion shaft bushes misplaced, and the best thing to do is to knock them out, and make new ones with the hole slightly eccentric. Reassemble and rotate them till you get reasonable gear-fit; then Loctite them in that position.

Trial Run.

Fig 4–5 shows a perspective of assembly. For the trial run, though, we do not need the drum or the reversing valve. Assemble the shafts and gears as shown, checking for correct clearances as you go. If the gears bind at the roots, lightly "top" the teeth by spinning at about 100 r.p.m. in the lathe and applying a very fine file.

Bed the faces of the cylinders and standard with a very little Brasso and oil, clean off, and assemble. Don't put too much tension on the springs. If the crank will not rotate, check that you have clearance each end of both cylinders; if not, adjust the position of the big end bearing. (Or, at worst, make a new piston rod!) She may be a bit stiff to rotate, but this will ease off after running a little. Stick a piece of pipe into each of the top holes in the standard, inject a few drops of oil, and apply compressed air. She will need perhaps 10 or 15 p.s.i. to start with but should soon buzz away merrily on 5 to 10. Look for faults, like rude noises, grinding gears, or grunts from the pistons, and correct each as it arises. Then fit the pipes to the opposite holes (a push fit should be quite good enough) and try her in reverse. When you have run an hour (which soon passes) take all apart, file little flats where the grub-screws were, thoroughly de-grease, and apply paint. This is up to you, but small children like bright colours, red cylinders, yellow gears, black winch plates (so that the oil doesn't show dirty), green jib, and all pipes etc., polished brass.

Winding Drum Fig. 4–6

This is shown as $\frac{5}{8}$ in. dia., but can be larger or smaller as you please. A smaller one lifts more, but carries less wire (string to you) and vice versa. It can be made of whatever comes to hand – even wood; boxwood will take a thread for the grub-screw provided it's not a fine pitch. I made mine of brass. You will see in the drawing that it has a thread on it. This is the wire-guide. Measure the cord (fishing line is ideal) and make the pitch of the thread just a trifle larger. Mine is 24 t.p.i. for cord 0.040 in. dia. Face the ends of the stock about 10 thou. less in length than the distance between bushes on the winch, and then turn a little shoulder to fit the $\frac{3}{8}$ in. hole in the side-cheeks. Drill and ream $\frac{3}{16}$

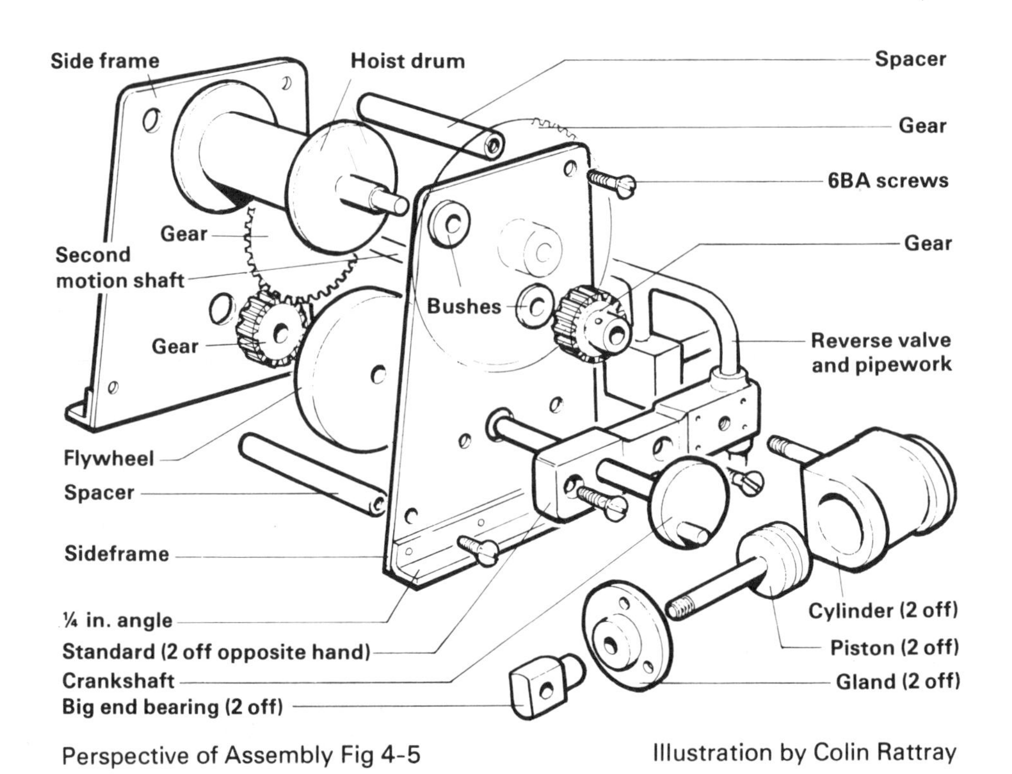

Perspective of Assembly Fig 4-5

Illustration by Colin Rattray

The completed winch and engine, with cord and hook.

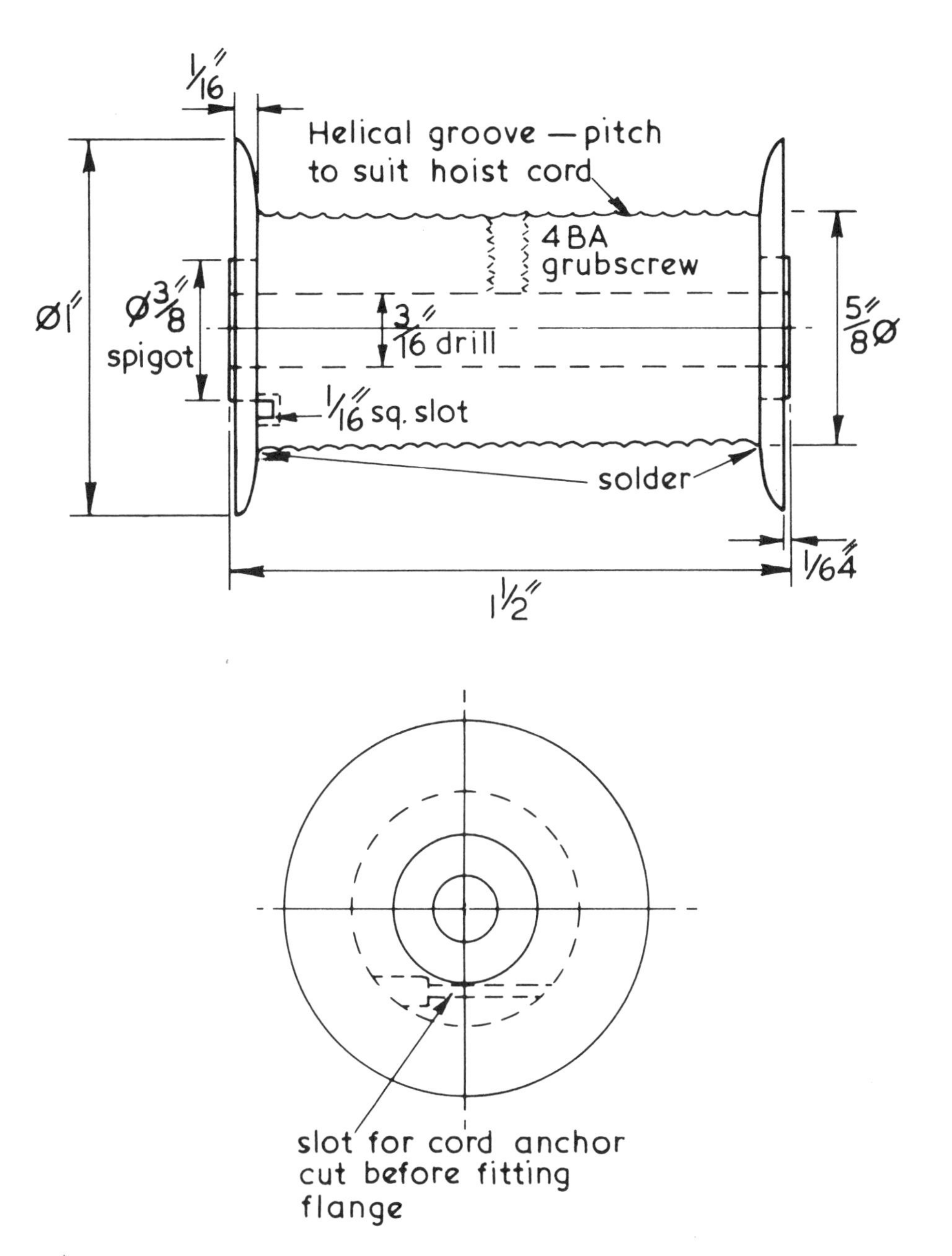

Winding Drum Fig 4-6

in., and remove sharp edges. The cheeks are parted off (or mine were) from 1 in. brass bar after drilling to fit the spigots, *but* before parting-off the faced end is filed slightly convex and *during* parting-off the cut is stopped and the edge well rounded. Make a saw-cut across the end of the drum, avoiding the bore, and enlarge one end to take a knot in the string. Then after soldering the cheek on you will have a hole through which the string can be passed. Check that you get the saw-cut at the right end, depending on whether you want to hoist with the rope on top of the drum or underneath. The first is "pukka", but it makes no odds. Solder on the cheeks, drill and tap for the grub-screw, polish all up, and there you are! File a flat on the shaft to match up with the grub. You can now fit it to the completed winch

Reversing Block Fig. 4–7

Make the block first. Face both sides in the 4-jaw, and then mark out for the ports, noting that two on one diagonal go only half-way through, two on the other right through. These are all drilled No. 41. Drill the centre-hole a close 6 BA clearance fit say No. 33. Turn over and very carefully enlarge the two ports showing on that side to half the thickness to be a close fit to the pipes you are using – $\frac{1}{8}$ in. or $\frac{5}{32}$ in. to choice ($\frac{1}{8}$ in. makes the pipework easier, but make sure it's fairly thin-walled stuff. Or you can use $\frac{5}{32}$ in. pipe with a $\frac{1}{8}$ in. stub soldered into the end). Now drill the access holes in the two opposite sides to meet the other pair of ports as shown on the drawing at "Section on XX".

The valve is faced in the lathe and the middle hole drilled from the tailstock. Mark out for four No. 41 holes, one at each end of the banana-shaped slots. These holes will, or should, match up with those in the block. Now make a little chisel out of 3/32 in. square silver steel and carve out the bananas. They don't need to be more than $\frac{1}{32}$ in. deep, nor need they be the exact shape, provided there is a reasonable flat surface between them and the centre hole or the edge. If by any mischance you mark the flat face, which might cause a steam leak, return to the lathe and reface. Once finished, part-off to thickness. (It's *far* easier to hold when making the grooves if you do this on the large chunk of stock.) Now slip a 6 BA bolt through and grind the two faces together, starting with the very finest emery, or even pumice powder, and finishing with rouge or Brasso. You should get a dull, not a polished surface.

Wash very thoroughly and then reassemble. Drill a $\frac{1}{16}$ in. hole in one corner of the block, and then fit a peg, preferably hard brass or stainless steel with well-rounded ends. Fit the valve, and mark on its periphery the position for a pair of radial pegs to limit the rotation – you find these limits by looking down the holes in the back of the block. Drill for and fit these pegs with the minimum of projection. The drawing shows $\frac{3}{32}$ in., but the smaller the better. Mine are just under $\frac{1}{16}$ in. Finally, hold the assembly in roughly its position at the back of the winch, between the engine standards; mark the top with letter "T" and also mark for the little 8 BA screw on the back of the valve to which the operating rod will fit. Drill and tap for this, taking care not to go right through the valve.

Pipefitting Fig. 4–7 and photo.

Make up the pipes as shown. Bend up the link-pipes first – those that join the cylinders – and offer them to the engine standards, adjusting till they look well and fit well. Make two short stub-pipes and fit these into the reversing block; hold this in place and make a mark on the link-pipes. Take all apart, file one end of each stub to a 90° point, and similarly file a 90° vee at the marks on the link-pipes. Make sure they fit closely, remove burrs and blow out with air. Clamp each stub to its link and braze up. Pickle, wash, and blow out.

Now offer these pipe assemblies to the

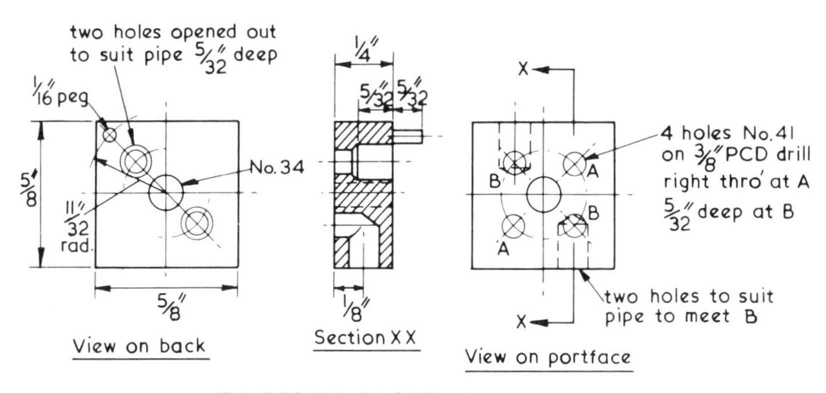

REVERSING BLOCK: brass

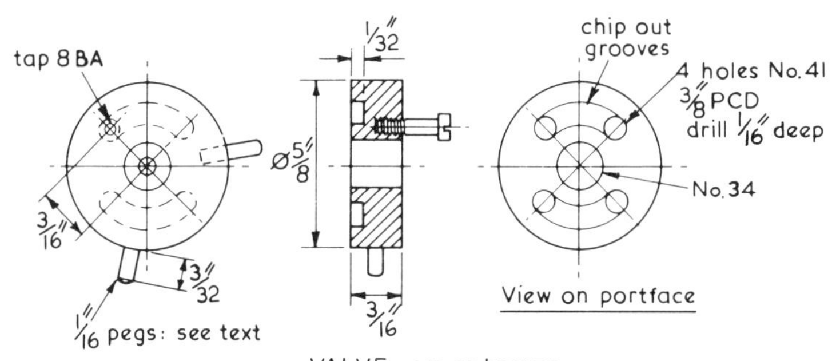

VALVE: g.m. or bronze

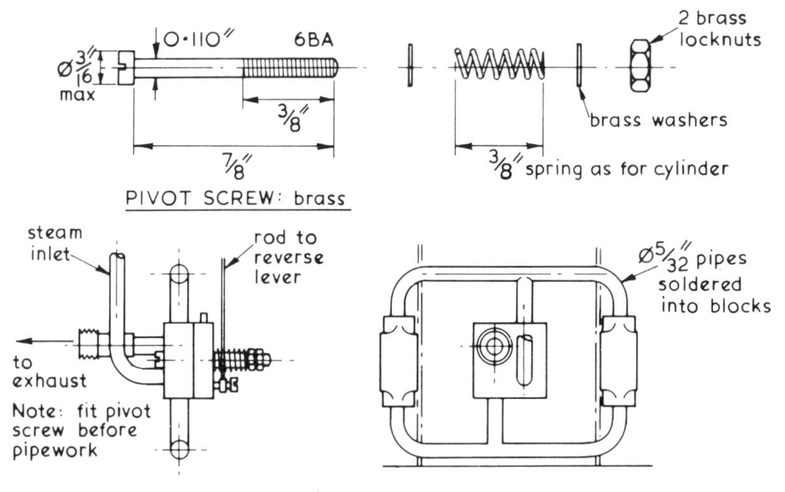

REVERSE VALVE & PIPEWORK

Fig 4-7

Close-up of pipework, reversing valve and lever, stop valve and lubricator.

job, and file the stubs so that the block sits square to the engine. Solder the upper link-pipe to the engine standards. Easy, just like that! The secret? Use a large soldering bit with a small end – I have a really hefty one with a $\frac{5}{16}$ in. hole in the end into which I can stick any shaped bit at will. Alternatively, use your normal soldering bit and help it out with a mouth blowpipe, or the fine flame of a Soudogaz lamp applied to the *pipe,* not the engine block. Get a good clean joint – it shows. Now solder the stub to the reversing block. Same procedure helps. You must then turn the job over and fit the lower pipe – doing the joints in the same order. Take care (a) not to get solder running *into* the pipes and (b) that the lower link-pipe doesn't stick out below the bottom edge of the winch plates. Clean everything well.

Fit a 6 BA round or cheesehead screw, put on the valve-disc with a spot of oil or grease, a washer, a spring like those used on the cylinders, another washer and a locking nut. Unlike the cylinder pivot spring, this one needs to be fairly well screwed up, as the steam pressure has a larger area to work on. You can now make and solder in the steam and exhaust connections to the back of the block, as shown on the drawing. Then, apply compressed air, or steam from a test boiler, and try her forward and reverse. If you *have* a little pot boiler available this is better than air at this stage as it will show up any leaks. Besides which, all engines run far better on steam than on air.

Reversing Lever.

I have given no dimensions for this as it offers an opportunity for artistic design on your part! You will see the way I did it in the photo. The little bracket is filed up from a bit of brass angle, drilled to pass the 6 BA spacer screw on one face, and drilled and tapped 8 BA for the lever pivot on the other. The lever is made from $\frac{3}{16}$ in. × $\frac{1}{16}$ in. steel strip, drilled about a third way along for the pivot screw and tapped 8 BA at the end. The operating link is bent up from a bit of 18 s.w.g. spring wire softened at the ends. Locknuts are used on the two little screws. You can make it much simpler, of course – just a bit of wire to pull up and down, through a hole in the winch frame spacer – or more elaborate if your tastes run that way.

That completes the winch, but I will mention a few odd points before going on to the next item. The $\frac{1}{2}$ in. pinion on the second motion shaft can be thinned down as there is no need for the $\frac{1}{4}$ in. face width on this job. I chucked mine by the boss and reduced it to $\frac{5}{32}$ in. wide on the toothed face. Use *two* grub-screws on the large gear on the winding drum; it has quite a torque to handle and one isn't really enough. (Meccano threads are $\frac{5}{32}$ in. Whit., by the way.) On the cylinder pivots and that for the reversing valve I used self-locking 6 BA nuts rather than locknuts. These came from Whistons at New Mills, and are very handy on jobs like this. The grub-screw in the winding drum is an Allen socket head type – an exception to my remarks about screwdrivers, but this one must be really tight. I think the photos should clear up any other queries.

Stop Valve and Lubricator Fig. 4–8

The reversing valve will work as a stop and start valve as well, but you need something a bit finer for speed control; and a lubricator is advisable. The design shown is more or less that due to the late "LBSC" of "Model Engineer" fame, and you will find that my description is almost the same as the old maestro's "words and music". I make no excuse – one of his that I made has worked perfectly for a quarter of a century; I only hope some of my designs last as well!

The body is made from $\frac{5}{16}$ in. square stock – $\frac{3}{8}$ in. A/F hex would do – the ends faced to $\frac{7}{8}$ in. overall length. I have a self-centring 4-jaw for this sort of work but you can set up in the usual way if you haven't one. Drill $\frac{7}{32}$ in. to $\frac{3}{8}$ in. deep, flat bottom the hole to $\frac{7}{16}$ in. and then re-centre with a little Slocumbe after which drill down No. 41 to $\frac{13}{16}$ in. total depth – just not quite breaking through. Tap $\frac{1}{4}$ in. x 40 using the tailstock drillchuck to guide the tap true. Remove from the chuck and drill $\frac{5}{32}$ in. holes for the spigots of the unions and the lubricator. Get them the right way round! The gland is made from a bit of brass hexagon. Chuck in the 3-jaw, face the end, turn down to $\frac{1}{4}$ in. dia. for $\frac{3}{16}$ in. and thread $\frac{1}{4}$ in. x 40 with the tailstock die-holder. Drill No. 30 and tap right through $\frac{5}{32}$ in. x 40. Reverse in the chuck and repeat. I use a spare union nut from my "bits" box for the gland nut. (If you do that, check first that it *is* 40 t.p.i. as some are 32; cut the thread to suit in that case.)

The spindle is stainless steel and should present no problems but use the tailstock dieholder and try to get a well-fitting thread. Make the cone on the end 60° for fine control. I silver soldered a brass boss on the end, cross drilled afterwards for the

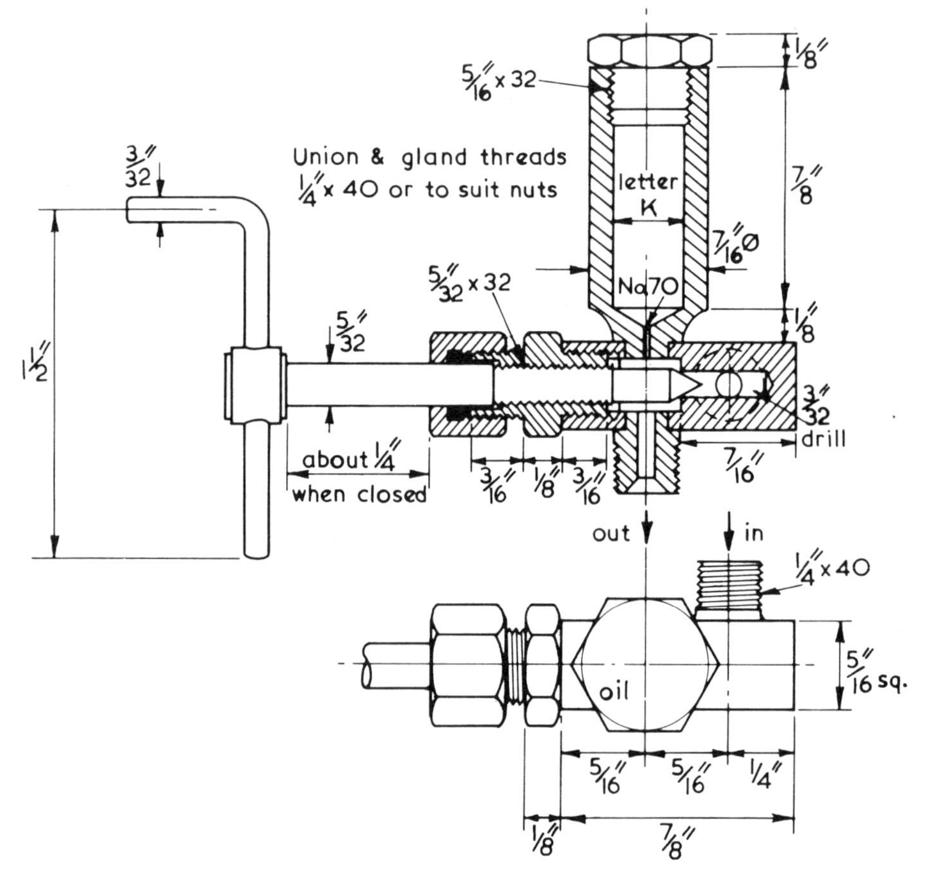

Stop Valve & Lubricator Fig 4–8

handle. Don't you do that! Cross drill the boss first, then drill and tap a $\frac{5}{32}$ in. x 40 thread, thread the end of the spindle, assemble all, and *then* braze (or even soft solder) the lot. The two union stubs should be threaded to suit the union nuts you are going to use; one of mine was 40 threads, the other 32. You can, of course, obtain "male" unions complete from Stuart Turners, Reeves, or Kennions, and cut them down if you like. But to make the stubs, simply thread a length of $\frac{1}{4}$ in. brass, drill $\frac{3}{32}$ in., apply a Slocumbe drill to form the cone, and part off. Reverse in the chuck, gripping very lightly, and form a $\frac{3}{16}$ in. dia. spigot.

The lubricator is a simple turning job from $\frac{7}{16}$ in. brass rod. The only snag is the small hole in the lower end. If you haven't a No. 70 drill (0.7mm.) use the smallest you have – say No. 60 – and burr the end a bit till a 24 or 26 s.w.g. wire just fits. The spigot into the body is $\frac{3}{16}$ in. dia.

The union bosses and the lubricator body are all three brazed in at one go, but take care again that they face the right way. Lubricator on top, unions pointing downwards and one to your *left* when looking *at* the handle end. Here a word about "handing" may be apposite. Quite a high proportion of the population is left-handed. If this applies to the youngster concerned, hand the whole winch – put the gears on the opposite sides to those shown, the reversing lever on the opposite side, and make the throttle valve opposite way round too. This brings the controls to the left-hand side of the crane (looking *at* the boiler end) and will be more natural for the driver when he (or she) stands behind the crane to work it.

There is no need for any support bracket, though you can make one if you like. The lower union stub is attached to the vertical pipe from the reversing valve; the other goes to the steamdryer pipe from the boiler, and these hold the job firm enough.

Boiler Fig 4–9.

This is $2\frac{1}{2}$ in. dia. x 4 in. long in the shell – you need a piece $4\frac{1}{4}$ in. long to allow for the flange. These sizes are not critical, and for jobs like this I tend to use the tube which is to hand or (more important) one for which I have flanging plates already! 20 s.w.g. is quite thick enough at this pressure (20 p.s.i.). The flue is $\frac{3}{4}$ in. copper water pipe, or the metric equivalent, and the end plates 18 or 20 s.w.g. sheet flanged to fit *inside* the shell. You will see from the photo that I have domed the top plate upwards as in real practice, but this isn't essential. You need bushes for the safety valve, water level plug and filler plug. I have made this separate from the safety valve, to reduce the risk of youngsters interfering with the setting. You can, of course, add fittings to your heart's content – level gauge, pressure gauge, and even a feed clack, pump, and water tank; but I think that the addition most likely to be appreciated is a whistle. This will not only use up surplus steam, but also serve to give warning to the cat that she and her basket are about to take a trip upwards! The bushes should be screwed $\frac{5}{16}$ in. x 32, which is the same thread as on the lubricator, so that the filling and level plugs can be identical. Further, this thread fits the Stuart and Reeves standard safety valves, if you decide to use ready-made ones.

In the introduction I promised not to refer back to previous pages, but at the same time I don't think I need go into quite so much detail on this boiler as the others – the procedure is the same. The end plates are identical – i.e. you need only one former turned to a diameter equal to the I.D. of the barrel, *less* twice the thickness of the endplates *less* another 5 or 10 thou. to allow for springback. A very small radius on the corner is needed. Having beaten the flanges down, set in the 3-jaw and trim them up a little, though as they don't show, extreme accuracy isn't needed. Mark out for the holes – see the plan view in Fig 4-

9 – and drill; for the centre hole I suggest with a ring of $\frac{1}{8}$ in. holes, then cut out, and finish bore it a tight fit to the flue in the lathe. Don't forget the $\frac{5}{32}$ in. hole for the steam pipe in the bottom plate.

I suggest you braze this one up in two stages, as the fit of the upper plate does affect the appearance. Nevertheless, flux *all* joint faces, and assemble the upper and lower plates, steampipe, flue and shell, but no bushes. The upper plate is there just to locate the flue. You will need a length of about 12 in. – more rather than less – of $\frac{5}{32}$ in. steam pipe, one end of which is bevelled as shown in the drawing. The other end should be squared off, but it will be brought to length when assembling later. With the work upside down, silver solder the bottom flange, steam pipe, and flue, then pickle and wash. Reflux, *including* the already brazed joint, this time fitting the bushes. Fit the top plate carefully so that the projection is even all round and, with the boiler upright, braze up the joints; touch the bushes with the alloy rod as you pass them and finally braze in the level plug bush in the side. Pickle again and wash out. Make sure that the pipe is well washed out.

Now check for soundness. Fill with water and look for obvious leaks. Any such at this "nil" pressure should be rebrazed, not forgetting to flux ALL joints when doing so. Next, fill with water and plug all apertures except one. Find a fitting which will screw into this, and attach to the cold water tap with hose. Inspect all over under tap pressure not forgetting to have the boiler both ways up, as it is unlikely that you will have got rid of all the air. Serious leaks must be rebrazed, but a mere weep can either be peened over with a ball-pein hammer, or, if it is not on the outside, caulk with a dab of soft solder.

Please note. In assembly do take care that the various bushes and the steam pipe emerge in the positions shown on the plan view of the drawing. It will be too late to alter afterwards. The actual steam pipe hole is more or less on the shell centreline, but after the pipe has been coiled round below (this is the superheater section) it can be bent to emerge about $\frac{3}{16}$ in. to one side.

Safety Valve. Fig 2–15, page 42

This is identical in design to that shown in Fig. 2–15 except that the body is larger and the thread is $\frac{5}{16}$ in. x 32. I know the boiler drawing shows a different one, simply because I used one I had by me. So, if you are going to make one instead of using a commercial product, proceed exactly as described on page 60, but use hexagon bar the same size as that for the lubricator cap, and screw the bottom end $\frac{5}{16}$ in. dia. x 32 tpi. In passing, some chaps have trouble making little screwed caps like this. It is quite easy. Chuck the bar and turn down the end to about 1 thou. less than the nominal diameter of the thread. Make a small bevel at the end (at about 60°) to assist the die in getting hold, and then make a 30° bevel on the face of the hexagon which does not quite reach down to the flats. With a narrow parting tool – about $1\frac{1}{2}$ threads wide – make a groove close to the hexagon which is a few thou. over $\frac{5}{8}$ x pitch deep – in this case $\frac{5}{8} \times \frac{1}{32}$ in. =0.020 in. deep. Apply the die from the tailstock dieholder and thread as far as you can. Reverse the die, but don't close it *quite* so much, and make a second run over the thread. Then part off, but before going right through put a bevel on the other face of the hexagon – I use my screwcutting tool for these bevels. In this way you get a good appearance, and a thread which will screw right in once the soft washer is in place. I use soft copper, made from sheet for permanent fittings, and leather for filler plugs at these low pressures.

Firebox. Fig 4–9.

The firebox is rolled up out of steel – I used

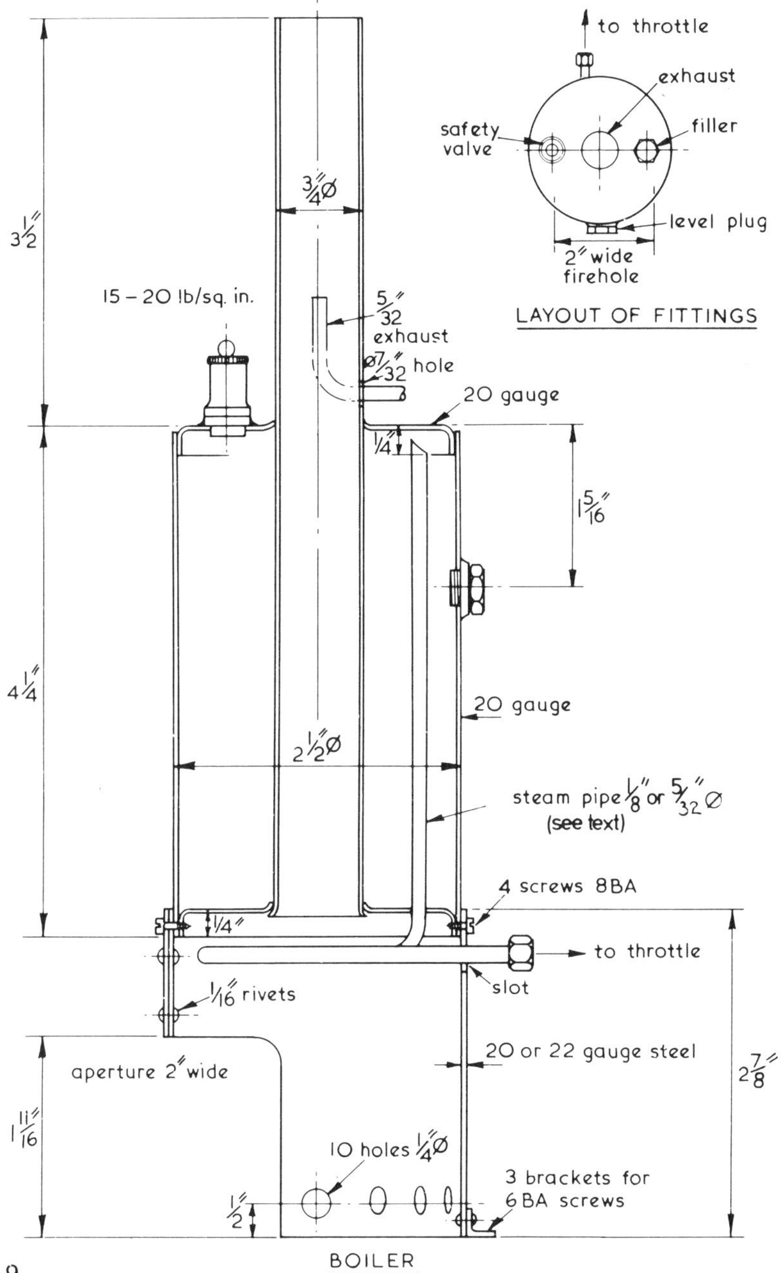

to throttle
exhaust
safety valve
filler
level plug
2″ wide firehole
LAYOUT OF FITTINGS
3½″
¾″ø
15 – 20 lb/sq. in.
5/32 exhaust
ø7/32″ hole
20 gauge
¼″
1 5/16″
4¼″
20 gauge
2½″ø
steam pipe ⅛″ or 5/32″ø
(see text)
4 screws 8BA
¼″
to throttle
slot
1/16″ rivets
20 or 22 gauge steel
aperture 2″ wide
2⅞″
1 11/16″
10 holes ¼″ø
3 brackets for 6BA screws
½″
BOILER

Fig 4-9

lead-coated "Ternplate" – or tinplate. *Don't* use aluminium; this is about as far as it could be from copper in the electro-chemical series, hence the corrosion experienced when aluminium washers are used on boiler fittings; an aluminium firebox would rot away in no time. You need a piece about $8\frac{1}{4}$ in. long to give the overlap. Note that the little cut-out to clear the steam-pipe is not exactly on the centreline – offer the rolled-up firebox to the winch and you will see where to cut it out to match up. Attach to the boiler shell with four 8 BA brass screws, and fit three little angle brackets to screw the boiler down to the baseplate. Offer up to the winch, cut off the end of the steam-pipe to length, countersink the end, and braze on the union nipple – *not* forgetting to put the nut on first! Braze in a $\frac{1}{8}$ in. stub if needed, as suggested earlier.

Lamp Fig 4–10

This is almost the same as that shown for "Polly" in Fig 2–15, except that in this case I squeezed the tube sideways to make it oval – about $1\frac{3}{4}$ in. wide x $2\frac{1}{4}$ in. long – so that it went into the firehole easily and the wick is larger. The little hat-on-a-handle is for damping down when idling on the engine. Don't forget the little hole in the top – without this spirit will tend to weep out from the wick. You will see that I have shown no filler plug; I find it just as easy to lift a wick out and fill that way. The metal of the body should be thinner, rather than thicker, and should be brazed, not soft soldered. See page 43 for procedure.

If you are going to use Meta fuels, make a sort of fence round a sieve. Use fairly coarse mesh wire, make the legs about $\frac{3}{8}$ in. tall and the fence round it about the same. It will need a little handle, and you'll have to buy or make a pair of tweezers so that the fireman can stoke to the back of the fire when need be.

Jib Fig 4–11

As I said at the beginning, this can be made to suit yourself and the contents of your scrap-box. A hexagonal or octagonal timber jib is quite attractive – there was just such a steam crane as this at a little country goods yard when I was a boy which had a timber jib; it was tapered at the ends and had a slight belly in the middle. The pulley axle and the jib pivot should be not less than $\frac{3}{16}$ in. dia. The pulleys on most "toy" cranes I have seen are far too small (that on mine may be a bit on the large side!) – there is a minimum size over which a wire may be bent. The groove should be U-shaped at the bottom, and the width of the groove about twice the cord diameter. The pulley axle is extended at either side of the jib-head to take the support wires.

These supports should be fairly stiff; $\frac{1}{16}$ in. welding rod will do very well. As you will see, mine are anchored back to the winch, but you could take them back to the rear of the baseplate if you liked. If you do this you may need a crossbrace at half-length to stop them flapping sideways. The bottom of the jib is supported in a couple of reasonably substantial brackets – whatever size of angle is handy. Say $\frac{3}{4}$ in. x $\frac{1}{8}$ in., cut down a bit in width where attached to the base. The "safe working load" is determined by finding what weight will tip the crane over and halving it.

Base Fig 4–11

The rotating base in my case is of $\frac{1}{8}$ in. steel. It looked a bit thin, and since the photo was taken I have rivetted some $\frac{1}{4}$ in. angle to it to give a more substantial appearance. Mark out the centreline first, and offer up the parts to this. Lay the jib flat along the centre and pop through for its holding-down screws; then the winch, and mark through for that; finally the boiler. Adjust the overall length of the plate to leave a bit sticking out behind the boiler. Now mark out for the centre pivot bolt – $\frac{5}{16}$ in. or $\frac{3}{8}$ in. – and make sure that its head won't foul the flywheel. (If it does, either

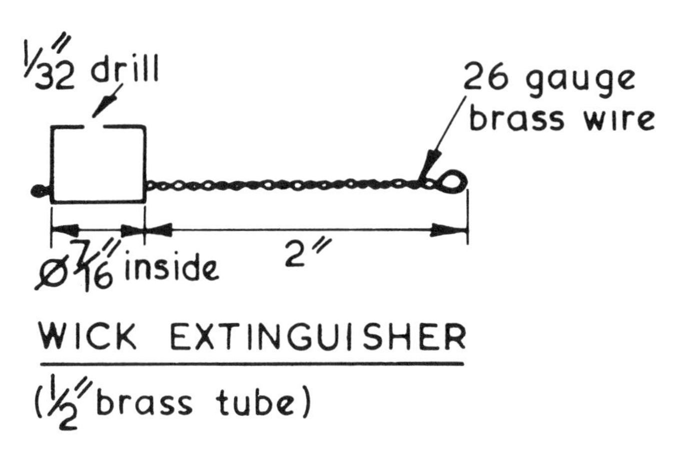

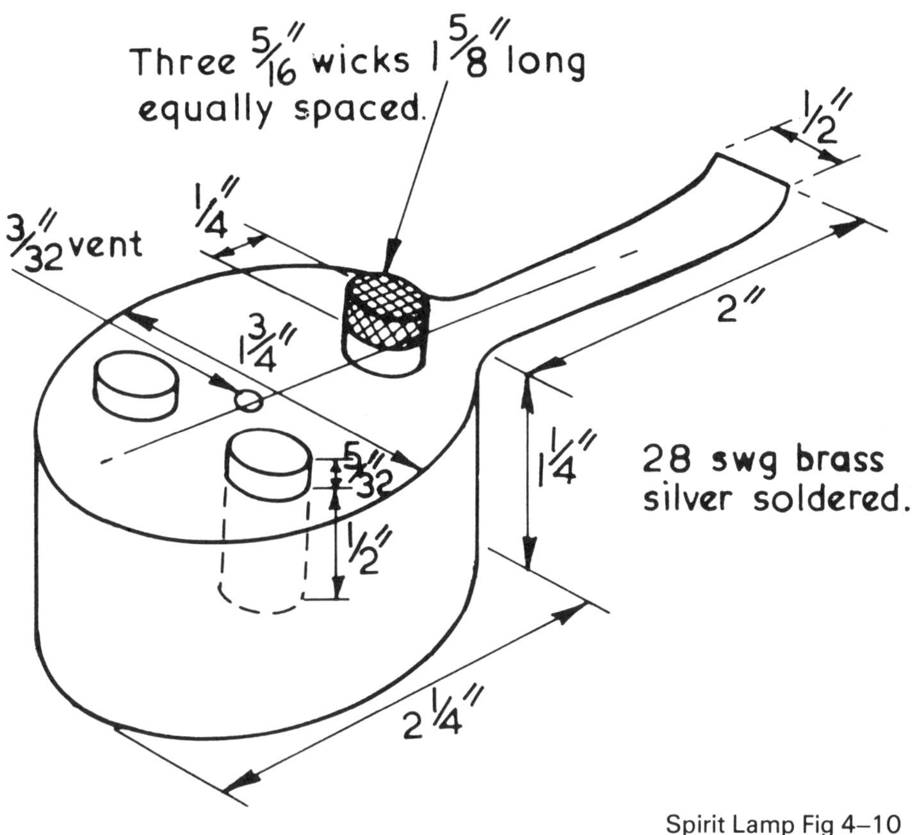

Spirit Lamp Fig 4–10

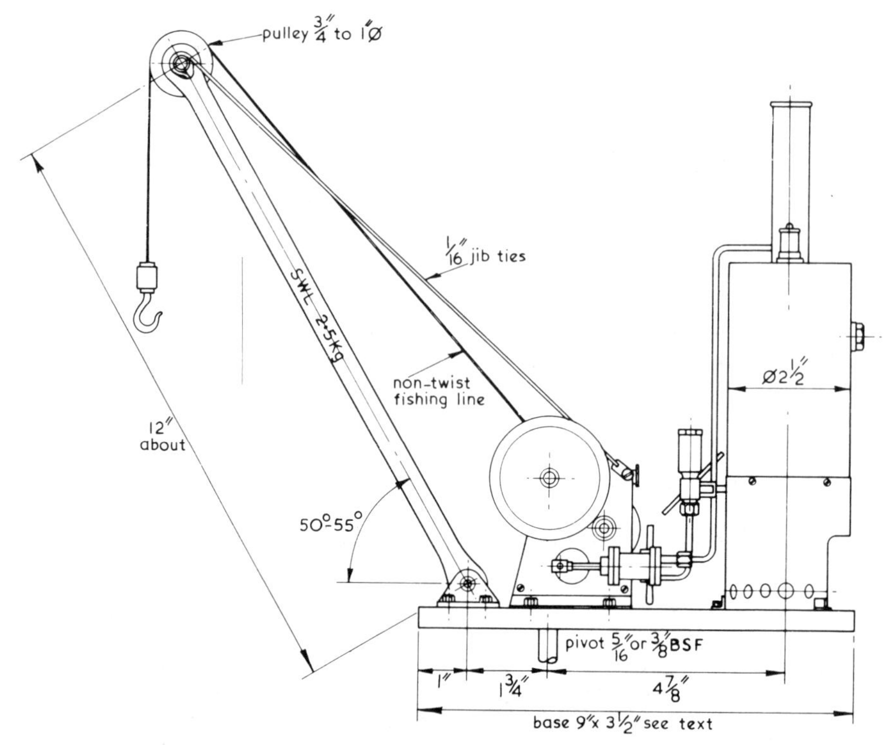

Approximate layout of parts Fig 4–11

reduce the head or move the wheel.) Drill through all fixing holes, and countersink below, or, if you are using wood, drill the appropriate pilot hole. Fit the pivot bolt before attaching the parts, and if you have turned off the hexagon I advise a hole across for a tommy-bar in the head.

Attach the winch, using countersunk bolts, and then the boiler. Hold all by the pivot bolt in the vice, and make any necessary adjustments to the steam pipe. There will be enough spring in the superheater length under the boiler to allow quite a lot of "wag" here, but I confess that the union nuts are a little difficult to get at.

Now for the exhaust pipe. This is led through the chimney mainly to reduce spillage of condensed water, but youngsters do like to see smoke coming out! However, there is a tendency to spit oil, as any condensation getting into the pipe tends to boil as soon as it gets into the hot flue and act as a steam gun. For this reason I suggest a little steam trap, which is not shown on the drawing.

Make a little pot out of say $\frac{5}{8}$ in. x $1\frac{1}{2}$ in. long copper tube with an 18 gauge bottom and lid. Solder a length of pipe into the side near the top with a union on the end to mate with that on the reversing block, and another in the top plate to fit the exhaust pipe to the flue. On the side at the bottom of the pot, opposite the entry pipe, solder a 4 BA brass nut, and drill through this say No. 33, and to this fit a 4 BA

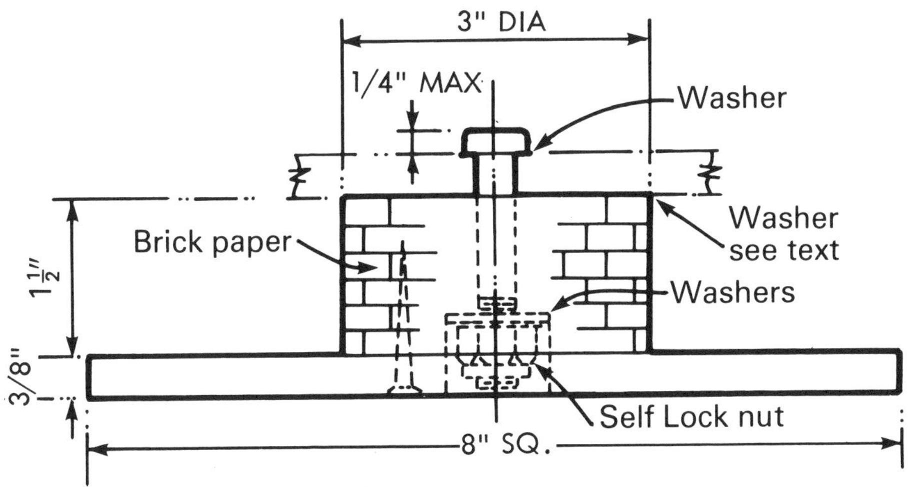

Suggested Wooden Base Fig 4-12

roundhead brass screw; this is the drain plug. The pot should sit happily on the base, between the winch and boiler, and the exhaust pipe can now be made and bent to fit into the hole in the chimney. It should point upwards and be as near the middle as you can manage, but there is no need to put a nozzle shape to the end; just cut off cleanly. You can, of course, fit a little cock to drain the pot if you like – which gives another piece of control gear for the young engineer to operate!

Attach the jib using countersunk screws again and thread the cord through the attachment hole in the drum. Use Meccano cord, or fishing line, whichever you please. Wind on one layer and see if this will give enough reach to satisfy you; there is, though, little point in having more cord than the height of a table. However, put on two layers if need be. Set up the stays, and adjust these if the angle doesn't look right.

Fixed Base or stand, Fig 4–12

The fixed base can be made to suit you – I give one suggestion. My own *was* to have been mounted on the loudspeaker from a defunct radiogram, which, being heavy and about 9 in. dia. on the frame, would have been ideal. However, looking for this in my junk-hole I fell over the base of a sun-lamp which came to us along with a box, several flower-pots and a few other odds for 10p at a local sale (we wanted the wooden box) and this is what I used. It looks quite well and is heavy enough not to tip over. The hole in the top was dead right, too! The one point to watch in this part is that (a) the crane can swivel sweetly and (b) there is no slobber in the joint so that the crane rocks. If you use wood I suggest a large diameter washer made of say ten-thou. shim-brass, and a spot of grease will help. Fig 12 shows one I made for a friend in America.

I see, I have forgotten to mention the hook. You can use the Meccano type if you like as it's quite strong, but you do need a fair weight on it so that the cord stays taut when reeling in with no load. I suggest a piece of $\frac{3}{8}$ in. brass, drilled through, and about $\frac{1}{2}$ in. long

General

The outfit needs completing with a little funnel for filling the boiler, a spanner to fit the plugs and the lubricator, a small bottle of oil and, I suggest, a tin which pours *properly* to keep the spirit in. This stuff is now sold in very dangerous plastic bottles, which not only disintegrate if they get hot, but also tend to squirt out the contents if squeezed *and,* moreover, don't pour. (The reason for forbidding the use of glass containers is *safety,* believe it or not!) You should also write out a proper specification and instruction sheet, as would the makers of a full-size crane.

You may be tempted to arrange the crane with a sheave on the hook, so giving twice the lift. This is up to you, but if you do you may find first that the thing twists (you will need a swivel hook to avoid this) *and* it halves the available hoist. However, the limit to the load is the "overturn moment" – it can't lift if it topples over! There is no reason why you shouldn't reel on two or even three layers of cord. The thread on the drum will reel the first layer quite smoothly, and the second and third will follow automatically. When the first layer is knobbly the others follow suit.

Some similar boilers are fitted with an asbestos liner to the firebox; I don't advise this for children's use; not a great risk, but I think parents would prefer that it wasn't there.

There it is, then. I think you'll find you need about a week of nights to make it, but it will give many years of pleasure. Indeed, if you make a really decent job of it you (or your descendants) may find it turn up at Christie's a hundred years hence as a "Contemporary 20th Century Toy Crane" and fetch a few dozen Megacredits, or whatever they use for money by then!

Chapter Five
JENNY WREN

A Miniature Vertical Steam Engine.

Here is an engine that is really out of the ordinary. A miniature working steam engine, which will run for quite long periods at 3000 to 4000 rpm on a couple of teaspoonfuls of water, almost dwarfed by a matchbox, and the base of which could, were it not for the law, be made from a 10-pence piece! (Older readers may think of using a "half-crown", but this is a bit too big!) It really does work, is more than attractive when set on the mantlepiece, and needs care rather than any skill at watchmaking in its manufacture.

You may think you need a "super precision" lathe – and it is true that mine was made mainly on a Boley watchmaker's machine – but this isn't so. The "UNIMAT" will fill the bill nicely, and provided the point on your tailstock centre is sharp, an ML7 should present no problems. What you DO need, though, is a good pair of tweezers and a little plastic box to keep the bits in. With a bore of about $\frac{1}{8}$ in. and stroke only $\frac{5}{32}$ in. the piston and rod are very small, and can easily be mislaid, as I have found several times. Another valuable accessory would be a jeweller's spade point drill in your archimedian (fretwork) drilling machine; if you haven't got one, ask your local jeweller to supply one, about $\frac{1}{2}$ mm point. (They *can* be had in sets, from 1/10 mm up to 2 mm by 5/100 mm steps too, quite reasonable in price.) You will also need a very sharp scriber, and I suggest you set a sewing needle in a wooden handle for the purpose.

Boiler, Fig. 5–1

My own boiler is made from a piece of thin-walled brass tube which originally formed part of a child's toy telescope. I have called for 26 gauge, but it can be 34 ($\frac{1}{4}$ mm) and be quite safe. Copper or brass to taste – brass polishes up better. Cut to length and fit wood plugs in the ends. Set up in the 4-jaw, gripping lightly, and tap the free end till it runs true, then set your tailstock to it. Trim the ends with a really sharp knife tool. Whilst set up, mark out centrelines – use a pencil, not a scriber – and set out the $\frac{3}{16}$ in. ventilation holes. If you put one on the centreline and the others at $\frac{1}{4}$ in. pitch, this will do. Mark out for the firedoor, too; it's easier when in the machine. Don't centrepop for the holes; apply your newly acquired spearpoint drill to the intersections of the lines and give a few strokes to the handle – this will give you an accurate centre, which you can enlarge with say a $\frac{1}{16}$ in. drill before following with the $\frac{3}{16}$ in. Which *don't* do till you have made the base.

You need formers for making the flanged top and bottom. Make that for the top first, just a circular piece of steel or brass, the end turned down to the O.D. of the tube less a few thou. Cut out a circle of brass or copper about $1\frac{1}{8}$ in. dia. and

Jenny Wren.

anneal it, then set to the former in your vice, with a backing plate about $\frac{7}{8}$ in. dia., and beat down the lip as described for the earlier engines in the book. (See page 22) Mark out for the holes, and use the former as a support to drill that for the safety valve bush – start with a $\frac{3}{32}$ in. drill and enlarge it by stages. Just make a popmark with your spearpoint for the steam-pipe at this stage – drill this when you have made the pipe. For the centre hole, drill it only $\frac{1}{8}$ in. at this stage.

Return the former to the lathe and reduce it to a diameter equal to the inside diameter of the shell tube, less twice the thickness of the endplate, and then about 0.005 in. smaller still. Repeat the process of beating down on the lower end plate until it almost enters the tube. Drill the centre hole $\frac{1}{8}$ in. as before. Now set this lower plate in the 4-jaw, gripping by the inside, and skim the flange till it just enters the tube, a push fit. Enlarge the centre hole by stages, and finally bore out a push fit to the piece of pipe you have for the chimney. This should be $\frac{5}{16}$ in. outside dia. *Note;* if you are working with a lathe of 3 in. centre height or larger you will find the chuck jaws too large to carry the plate. In this case, chuck a piece of hardwood – boxwood is the ideal – and turn a spigot on the face to carry the workpiece. If it tends to slip use a little piece of double-sided tape to hold it. Or you can make some shellac cement by dissolving shellac in methylated spirit. Paint this on the work, apply to the spigot, and as soon as it has set you can proceed. Remove the work either with a sharp tap, or by soaking in meths. Repeat for the upper plate.

Having done this, hold the plates, in turn, in the 3-jaw, flange outwards, and skim this to truth. Bring both plates to a smooth finish with fine emery; polish out any scratches, too, as these will be difficult to remove later. Set the plates aside for the present.

The base is a simple turning job, and by no means critical to shape or dimension. The only thing that matters is the spigot which fits the tube. Having made it, fit it to the barrel and insert the lower plate at the other end. You can now support the barrel on a vee-block whilst drilling the ventilation holes. Cut out the firehole, too. I recommend you drill a $\frac{1}{16}$ in. hole at the corners of this aperture, and file or saw to the corners. You must go carefully, or you will bend the thin material. I don't advise using tinsnips – the tool I use is an Eclipse No. 45 "Backsaw", the finest blade of which has so many teeth/inch that you need a glass to see them. A fine piercing saw would do. Clean up the outside of the barrel, but retain a faint mark to show the centreline at the top.

The chimney is made from $\frac{5}{16}$ in. OD tube, 22 or 24 gauge; In my case I had none thin enough, so I used a piece of thickwalled brass tube which was in fact 7 mm bore, and turned the O.D. down to size. Don't try to drill out thick tube, for the drill will almost certainly seize in the bore. Use a small ball-pein hammer to beat out the bell mouth at the top which, when formed, you then trim in the lathe. Trim the other end likewise, and remove burrs.

The bush for the safety valve is a simple turning job, though small. You need a very sharp knife tool and an equally sharp parting tool. Part-form the thread with the taper tap, but don't forget to open it out after the brazing operation. Make the spigot diameter a good fit to the hole in the top plate.

Assemble all, with flux at the joints, (but leave off the base; this is soft soldered in place later) the bush for the safety valve being aligned with the centre of the firehole. Set the assembly upside down in the brazing spot and braze up the two plate flanges and the lower flue joint. Use Easyflo, as this gives nicer fillets than the No. 2 alloy. Turn the work over and braze the upper flue joint and the bush. Be sparing with the alloy – use enough to

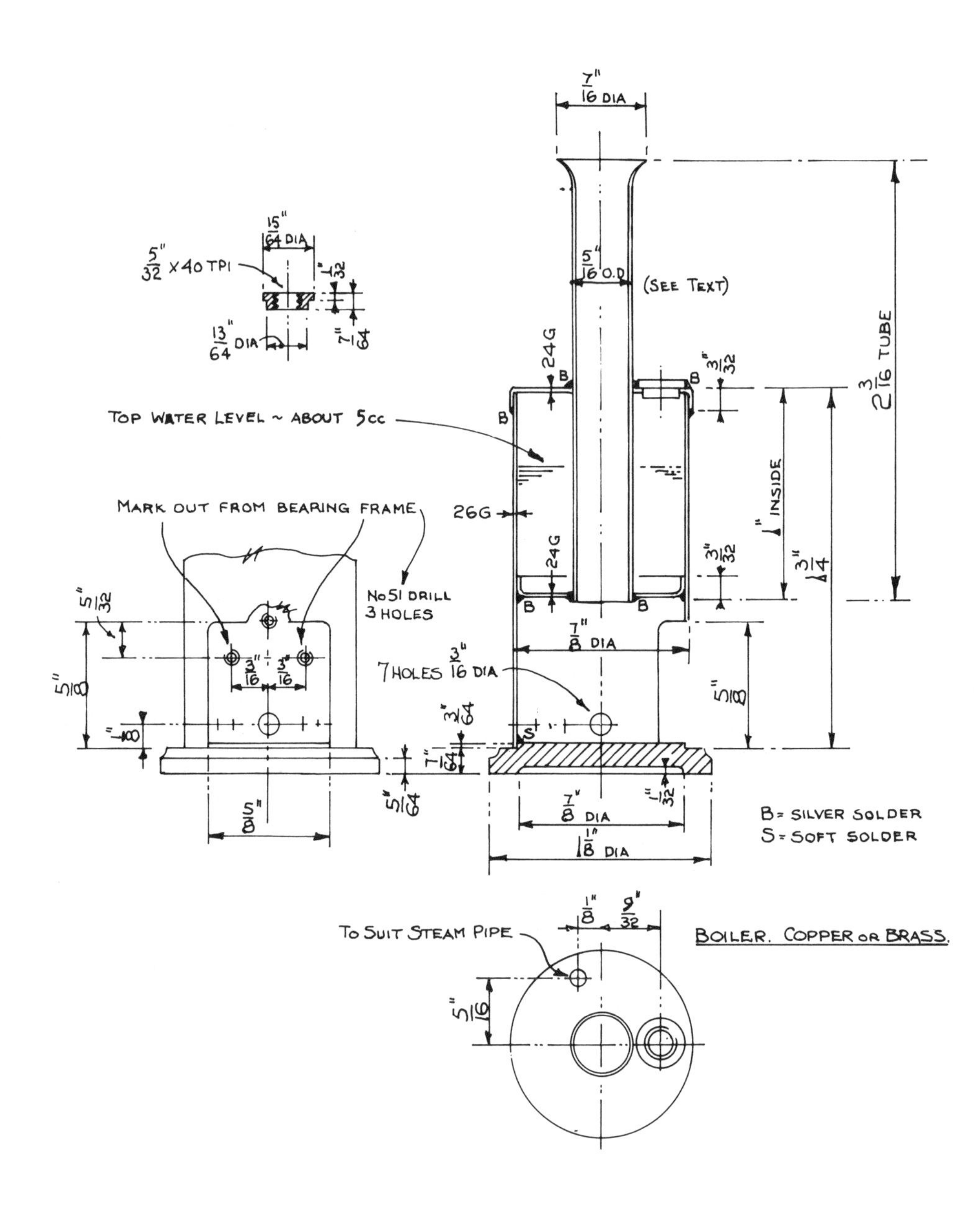

Fig 5–1. Boiler.

make a good joint but no more. Pickle well, with the boiler upright so that the acid gets inside, and wash out very well indeed. I recommend you to boil the job in a saucepan after the final rinse, to get rid of all traces of flux. Test for leaks as described in Chapter 2. Clean up the exterior and tap out the bush thread.

Tin the inside of the tube bottom, and the spigot of the base, and soft solder these together; again using the minimum of solder – all inside, so that it doesn't show. You must now set out for the upper *only* of the three holes which will carry the engine frame. The others are marked out from the frame itself. Take care to get it on the centreline and drill No. 50.

You can now give the whole issue a good polish, and set aside for the time being.

Cylinder. Fig. 5–2

First let me say a word about this class of machining generally. The secret is to use very sharp tools indeed, set exactly at centre height, and to take cuts of only a few thou. in most cases. This means that your knife tool – the principal cutting tool you will use – must have a *point* on the end. Having ground the tool to the correct angles, stone it with a fine india oilstone till no grinding marks show at all. Then finish the surfaces at the cutting edges with a hard Arkansas Stone, which will give almost a polished surface. Finally, give just *one* stroke of the Arkansas across the very point of the tool, just one, no more. Keep the tool in this condition throughout. Give similar treatment to any other tools you may use.

With these small workpieces you can – and must – run the lathe fast; You may have to go a bit slower when dealing with the slender parts in stainless steel, but even here I was running at 1000 rpm most of the time. Cuts will be very fine and so will feeds. You may need to rig up a magnifying glass to see what is going on!

The cylinder barrel is made from $\frac{5}{32}$ in. or $\frac{3}{16}$ in. dia. brass stock. Use a piece about $1\frac{1}{2}$ in. long; face the end and centre with a small Slocumbe. Draw out about $\frac{5}{8}$ in. from the chuck and drill No. 32, $\frac{11}{32}$ in. deep. To bore out, make a little boring tool from $\frac{1}{16}$ in. silver steel. Alternatively, find a broken HSS drill $\frac{1}{16}$ in. dia. and grind the end flat across. Grind about $\frac{1}{16}$ in. or $\frac{3}{32}$ in. of the end down to half the diameter, and finally form a small relief. If this is set in a holder– just soft solder it into a piece of say $\frac{1}{4}$ in. square steel with about $\frac{3}{8}$ in. projecting – and arranged in the toolpost at a slight angle it will make a most effective boring tool for this job. You must stone the end, as described above, of course. Run as fast as you can without chatter and take out no more than a thou. at a time, to clean up the bore to run true. Once this is reached, take out one more cut, with a very fine feed indeed. The exact diameter doesn't matter, as you will make the piston to fit. You can estimate the dimension of the bore later by turning a slightly tapered test piece, entering this to the bore (without force) and measuring the point to which it enters.

If you have never worked on such small parts before you must expect to have some difficulty, especially with seeing what you are doing. Don't worry; it will pay to make one barrel as a practice piece, and then make another "for real". Once bored, turn the O.D. to final dimension. Mine is 9 thou. thick. Don't part off yet, but make a little groove at the $\frac{3}{8}$ in. dimension; you will need the surplus within the chuck to hold the barrel when you lap the piston.

For the portblock I have suggested you drill a block of brass and later file to shape. The drawing shows No. 27, but you can alter this to suit the final O.D. of the barrel if you have adjusted any dimensions. Having drilled the block, set on a mandrel to face the ends and then file to the $\frac{7}{64}$ in. figure. It is important that this face be parallel to the centreline of the hole, so check with the mandrel on a surface plate.

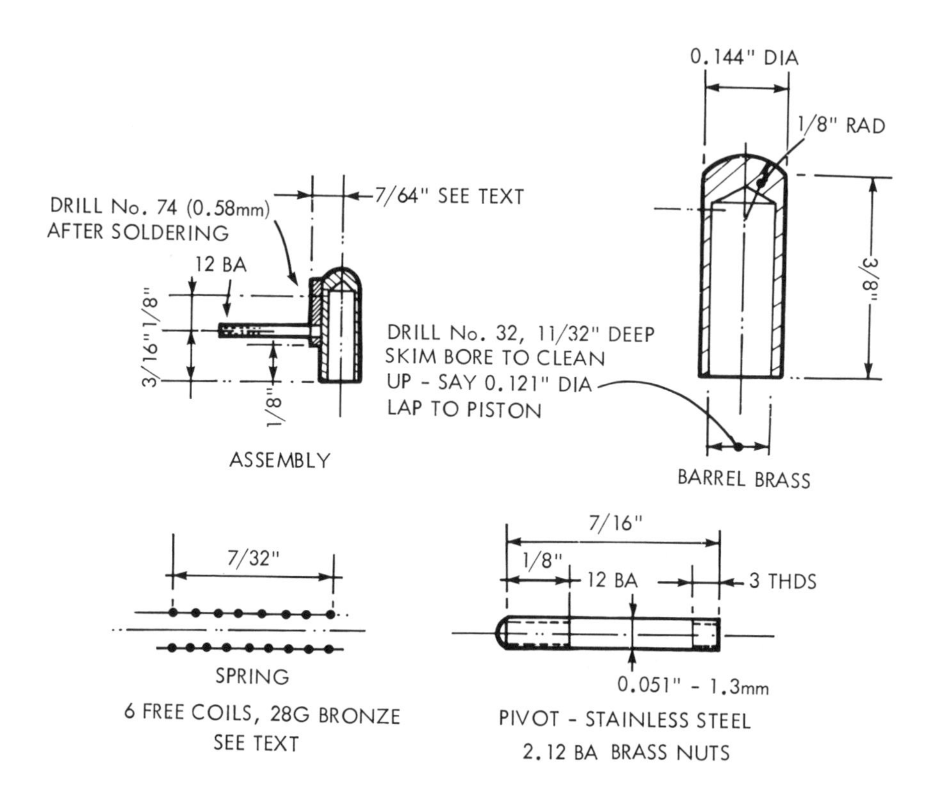

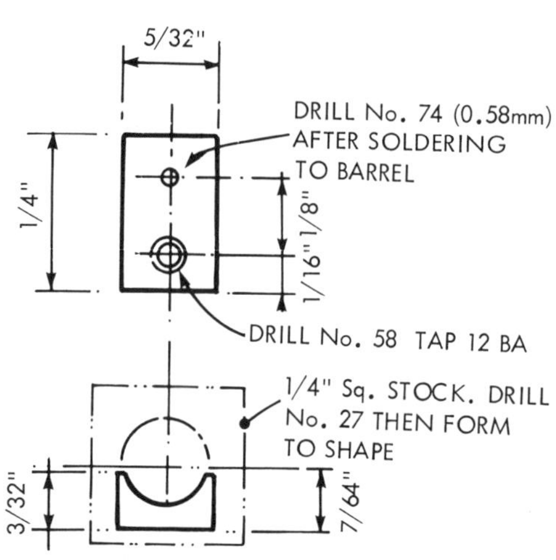

Cylinder details Fig 5-2 PORTBLOCK - BRASS

Still on the mandrel, set up in the lathe with the filed surface vertical and mark out the centreline with your needle scriber. Set out the centre of the 12 BA hole and make a pop with the spearpoint drill. From this centre, very carefully mark for the $\frac{1}{8}$ in. distance to the steam port, and pop this with the drill. You can now drill these holes whilst the block is still easy to hold; I know the drawing says drill the No. 74 after soldering, but I suggest you drill the block a size smaller – No. 75 or 76 – now, and use the correct size to go right through after soldering. Finally, after tapping the hole, which you must do very carefully to keep square, cut off the surplus and form to size.

Tin the faces of block & cylinder and wipe off any surplus solder. (The barrel still has its stalk aboard to hold it with) reflux and hold the two parts together with a binding of black, thin, iron wire; if you have none, just heat some bright wire to red, to oxidise it. Adjust carefully to the position shown in the assembly sketch. Now heat the *stalk* (not the work itself) and apply just a little cored solder to the work – just a very little. You must not run the risk of getting any solder into the tapped hole, as it would be most difficult to remove. In fact, the tinning should be enough to sweat the parts together but additional solder may be needed if the hole in the block doesn't exactly fit the barrel. Clean off all flux. You can now drill the No. 74 right through into the bore. Check that the portface is parallel to the bore, and if not, correct it with a very fine file; finally flatten the face by lapping to a piece of glass – mirror glass, not the window sort – with metal polish.

The pivot pin should present no problem; I have called for stainless steel, but in fact hard brass will serve. For the threads use your tailstock die-holder, and note that there are only three threads on the end that fits the cylinder; any more, and there is a risk of denting the bore when screwing in. I took a great deal of trouble over these threads, to get them tight but at the same time avoiding distortion of the bore. Today you can use Loctite to secure the spindle; much easier!

The spring will need some experiment You need just enough pressure to hold the parts together without introducing too much friction. I have suggested a size on the drawing – the same as mine, which was wound on a $\frac{3}{64}$ in. silver steel mandrel – but I would recommend that you try one of 30 gauge or even thinner first. It is important that the ends be ground square, so wind on a few more coils to allow for this operation. I should have told you to polish the pivot pin, and especially remove burrs at the thread ends. With a tiny cylinder like this all excess of friction must be avoided. Set aside till you have made the piston, but I will tell you now how to finish the job off (*after* lapping the pair to each other).

This job is done in effect by a combination of parting off and form-tool. Grind the right-hand face of a parting tool to approximately $\frac{1}{8}$ in. concave radius, and stone this with a gouge slip till all grinding marks have gone. Reduce the width of the tool to say $\frac{1}{32}$ in. at the point, or even less. Check the shape of the tool to ensure that when the point reaches the centre of the work the portblock will clear its shank. Polish the top face of the tool – you need no top rake. Set exactly at centre height. Chuck the work by the stalk with sufficient projecting to allow about $\frac{1}{16}$ in. clearance between tool and chuck, and in this case running much slower – say about 250 rpm – carefully part off, with a bit of wood (a match) to catch the work, held inside the bore. You may get a little pip on the top of the barrel, but this can be cleaned off with a No. 4 Swiss file.

The alternative method is to part off in the ordinary way, and then reverse in the chuck to form the end, I think you will appreciate, though, that there is grave risk of distorting the cylinder when holding it. It *can* be done, but the risk is not worth taking. Another way is to form the top of the cylinder before drilling and doing the rest of the work, but unless you have a watchmaker's lathe with collets it will be difficult and, moreover, you will have problems both with the soldering and the lapping; when you have it finished you will see what I mean – it is somewhat small!

Piston & Connecting Rod Fig. 5–3.

This is made from free-cutting stainless steel. (EN58M or 303S21 is the specification if you have to buy any) An alternative would be drawn gunmetal, but don't use brass or phosphor bronze. Having a lathe with a chuck that *is* true, and collets as well, I started with $\frac{1}{8}$ in. dia., but better to use $\frac{3}{16}$ in. if yours is the normal chuck. If you have a small lathe (e.g. Unimat etc) it is worth using the tailstock to support the work so face the end and put in a tiny centrehole. Otherwise you must work with the stock projecting free about $\frac{3}{4}$ in. from the chuck. Make sure your tools are sharp as previously described – you will need one with a very small radius on the end as well as the knife tool. Run at about 1000/1200 rpm unless you have trouble with chatter.

Start by roughing down all with the knife tool till you are a few thou. oversize to the piston diameter. Check that you are turning parallel at this stage and correct any discrepancy. Then finish the diameter till it just enters the cylinder – a pushfit, not a running one. Form the three grooves at the end nearest the chuck – they need only be 10 thou. wide by 6 thou. deep; the point of your knife tool will serve. This may form burrs. Remove them by dropping the speed and holding the thin end of a fine round Swiss file in the groove – take care that the file doesn't slip sideways. Make a somewhat deeper groove to indicate the top of the piston and then take all measurements from this. Face to length at $\frac{9}{16}$ in. + $\frac{3}{64}$ in. from the top.

Now turn down the end to $\frac{3}{32}$ in. dia., over a length of about $\frac{3}{16}$ in. and with a fine finish. Put a slight bevel on the free end. Measure this diameter, and calculate the infeed required to go down to $\frac{3}{64}$ in. dia. It should be about 23 thou. With your round-nose tool feed in this amount, working the tool a little from side to side, to form the radius at the "big end" – the actual size of this radius is unimportant; $\frac{1}{16}$ in. is about right. You now have two alternative procedures. You can carry on with this tool, reducing the shank a couple of thou. at a time full length, and so automatically forming the radius at the piston end; or you can use the knife tool, leaving about $\frac{1}{16}$ in. at the piston end for final forming of that radius. The advantage of the knife tool is that it puts very little bending load in the work, but you must be sure that it is really sharp, right to the point. The round-nose will give you a better finish, but as you will see when you get near to size, $\frac{3}{64}$ in. dia. is pretty small! The risk of bending it is greater. I used the knife-tool method and with a very fine feed, about $\frac{1}{2}$ thou. cut, and high speed; the finish was quite acceptable.

Let me repeat; if you are unaccustomed to this scale of work don't regard the spoiling of the first attempt as a "failure" on your part. The second one will be much better, and if you have to make a third to be satisfied, why not? You are doing this for fun, not to make a profit – and even those who *do* do this class of work for a living usually make a few trial pieces first. Don't be reluctant to take time out to rehone the tool cutting edges, either. The tool must cut; if it once starts rubbing, especially on stainless steel, the job will be spoiled.

You can now part off about ten thou. overlength. Reverse the piston in the chuck (don't drop it in the swarf tray or you will surely lose it) and face the end. Get a good polish with fine emery; this will reduce condensation losses. But keep it flat.

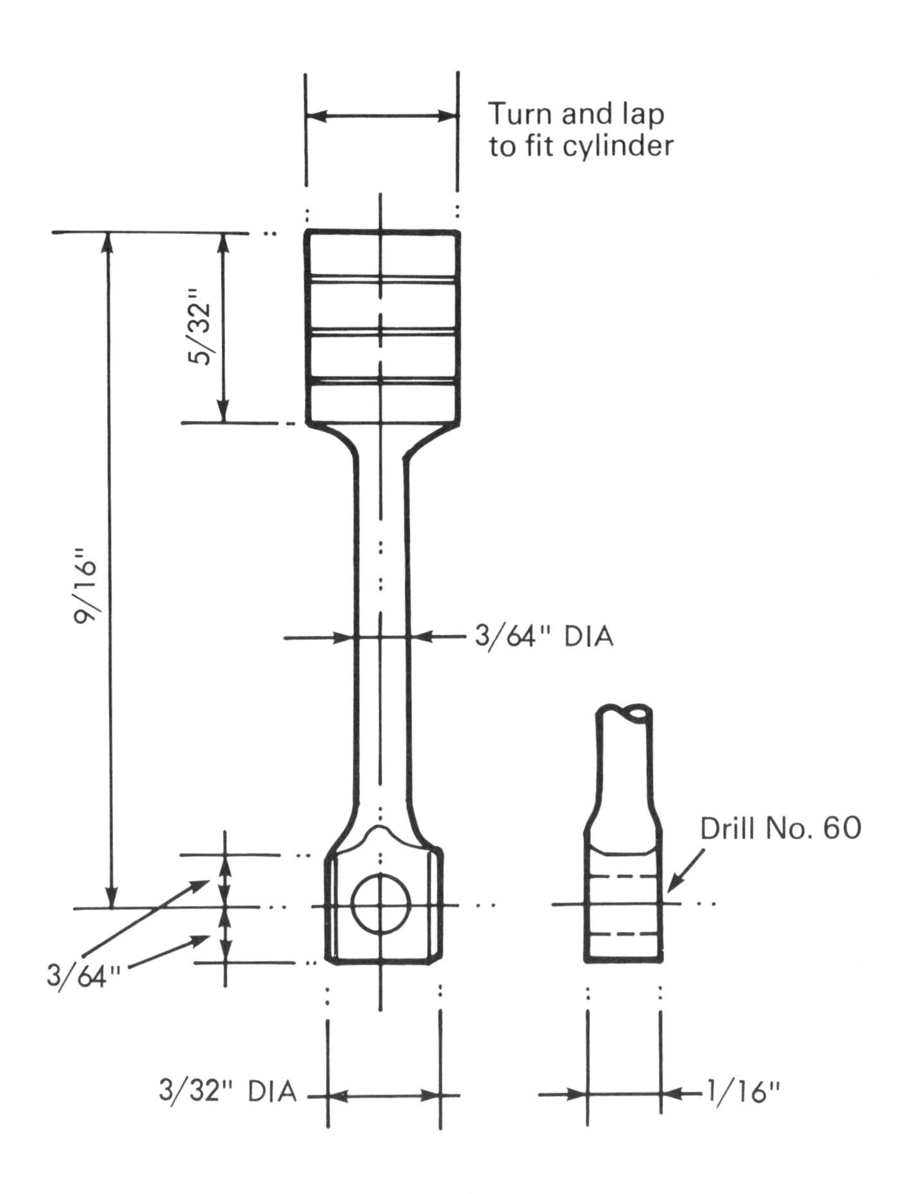

Piston and Connecting Rod Fig 5-3

For the lapping operation, hold the cylinder assembly in the chuck by its stalk and set the lathe to run about 40–50 rpm. Apply a little metal polish and with the lathe stationary essay a few strokes up and down with the piston. If there is a tight spot, give a few more rubs just over this length. Remove the piston, clean, and apply more polish. This time run the lathe and, holding the big end in your fingers – still a round knob – run the piston back and forth at somewhat less than the lathe speed, the piston running out about $\frac{1}{32}$ in. from the bore. (The bottom edge of the piston, that is.) After a minute or so clean off all polish from piston and bore and try the fit. Don't force it if it is still tight; just keep on with the lapping, cleaning and recharging with polish as you go. If the piston offers to "seize up", let go immediately. Don't try to pull it out – warm the cylinder with a spirit lamp and the piston will come free. At all times use a delicate touch and force it not.

When it is "cooked" the fit will be an easy push fit when dry and a close push fit with some light oil on it. This means it will be a nice running fit when the whole is at steam temperature, as the brass expands more than the piston. Don't go beyond this stage, as the engine will run in once it starts working; and in any case it is easier to ease a tight fit than to rectify one which is too sloppy.

Once this is done, file the flats on the big end. The hole can then be marked out, noting that the $\frac{9}{16}$ in. figure is important. Set the work up standing on the piston head on the lathe bed and use a carefully set scribing block or a height gauge to scribe the line. Get it central to the big end by eye, pop with the spade drill and then drill No. 60. Take care to get the hole square.

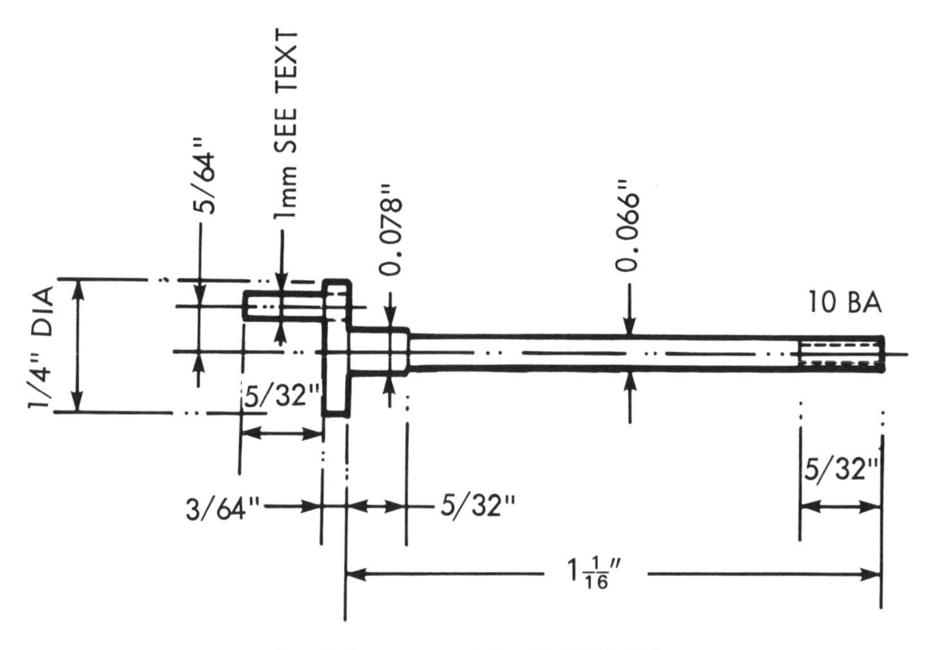

Fig 5–4 Crankshaft   Stainless steel (or F.C.M.S.)

You can now part off the stalk from the cylinder assembly as previously described.

Crankshaft Fig. 5–4.

The shaft itself is turned from a piece of $\frac{1}{4}$ in. dia. stainless steel, and the crankpin forced in. You will see I have called for this to be 1 mm dia. This is because I used a piece of "blue clock-pin steel" for it. You should be able to get this from any watch repairer, or you can buy it in a packet of assorted sizes quite cheaply from A.G. Thorntons of Heaton Road, Bradford, Walsh's of 12 Clerkenwell Road London, or any horological material supplier. It is very useful stuff to have about the place, being carbon tool steel, hardened and tempered to blue – it is JUST machineable, yet hard enough to stand wear and very strong. If you can't get it, then use a bit of 1 mm silver steel, unhardened.

Chuck a piece of $\frac{1}{4}$ in. dia. stainless steel, the free machining type preferably, about 2 in. long and polish the O.D. Rechuck with $\frac{3}{16}$ in. projecting, reduce this to 0.066 in. dia. for 10 BA and form the thread with your tailstock dieholder $\frac{5}{32}$ in. long; that is $11\frac{1}{2}$ threads. Draw out another $\frac{3}{16}$ in. and turn this down to 0.068 in. – 2 thou larger. Repeat at this dimension again. You will now have the thread, plus about $\frac{3}{8}$ in. at 0.068 in. Remachine this down to 0.066 in. to make a good finish, as this section is a bearing. It won't matter if it is a thou. oversize as we can make the hole to suit, but it does need to be a good finish, polished with fine emery if need be.

The rest of the shaft at this size is then machined in exactly the same way, $\frac{3}{16}$ in. at a time drawn from the chuck, and turned down with the knife tool. The finish need not be so good here and the diameter isn't important (so long as it is between 0.066 and 0.070) as it is a clearance in its hole. The final $\frac{5}{32}$ in. of this $1\frac{1}{16}$ in. length is machined to a fine finish at 0.078 in. dia. – this too is a bearing surface. Part off a little over the $\frac{3}{64}$ in. thickness and then reverse in the chuck to face the crankdisc to dimension and a fine tool finish. Whilst in the lathe, use your scribing block to mark out the $\frac{5}{64}$ in. radius to the pin hole. Centre this with your spear-point drill, and then set up truly vertical and drill No. 61. Cut off the pin to $\frac{13}{64}$ in. long, round one end and slightly taper the other to press into the hole. Again, these days you can drill 1 mm if you like, and secure the pin with Loctite.

The procedure so far outlined assumes you have a chuck which runs true. In fact, even if it runs out a bit it won't matter much provided you don't *rotate* the stock as you draw it from the chuck; the centre of the shaft will still run true and it doesn't matter a bit if the $\frac{1}{4}$ in. O.D. is not. However, you may have a spot of bother if your chuck is bell-mouthed. In which case, wrap a piece of paper round the stock at the mouth of the jaws. It also helps if you experiment to see if using the chuck key in a different hole pulls it more true. Most chucks have one keyhole which gives the truest running work. In extreme cases, make a "false collet". Chuck a piece of $\frac{3}{8}$ in. dia. brass, drill and bore this to $\frac{1}{4}$ in. dia. Mark the position of No. 1 Jaw, and use No. 1 keyhole. Take out and split about 15° from the jaw mark, replace with this mark against No. 1 jaw, and thereafter don't move it. It will grip well enough for your purpose, and though not perfect may well give a truer job. Use the No. 1 key-hole every time.

Flywheel Fig. 5–5

This requires little comment. It can be of brass or steel, and is a straight turning job. However, it is desirable that the "wheel" and the hole be done at the same setting so machine the boss and the recess on that side first, then turn round in the chuck, turn the O.D., face and recess, then drill as indicated on the drawing. Note that the boss projects $\frac{1}{64}$ in. from this face. I have shown a little belt groove in the boss, but I doubt if the engine would drive anything!

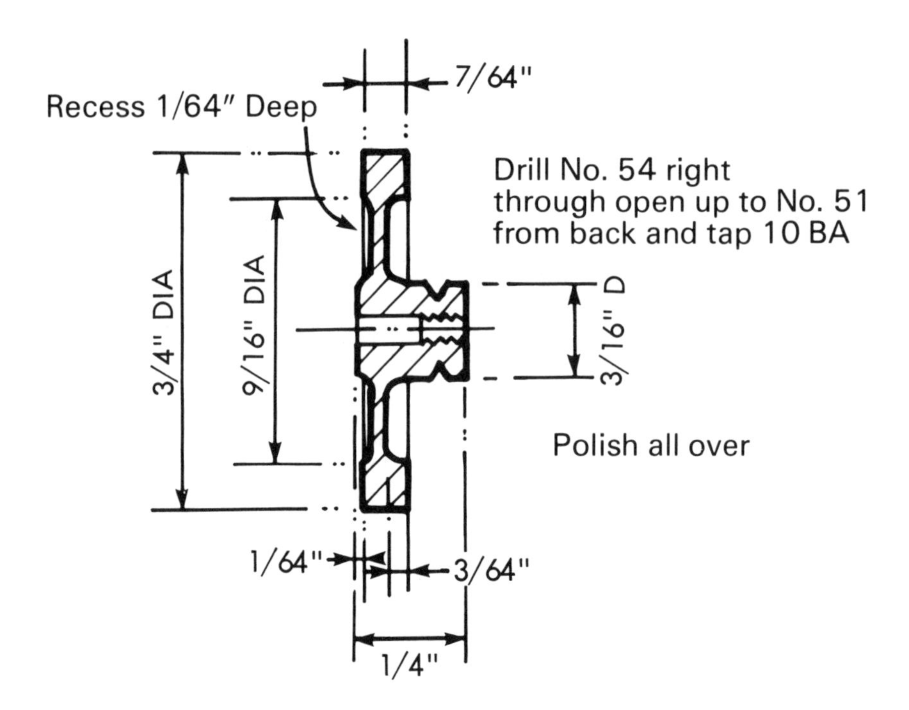

Fig 5–5 Flywheel

Not so much on account of lack of power, but simply because it isn't heavy enough to stand the pull of even a very light rubber band.

Standard, Fig. 5–6

This is mainly a careful marking-out and drilling job. Run over the stock with a fine file to clean it up; don't polish with emery at this stage. Use a surface gauge and plate to set out the main centreline with your needle scriber and pop for the lower hole. Set your dividers to $\frac{9}{16}$ in. and mark, then drill-pop for the No. 54. The $\frac{1}{8}$ in. radius line should present little difficulty, if you use an eyeglass to set the dividers, but the two 0.025 in. lines may be difficult.The way to do it is this. When you make the longitudinal centreline the scribing block and the work will both be resting on your surface plate or lathe bed. If you now set a .025 in. strip under the work this will centre it correctly for one hole, whilst if it is set under the scribing block this will set for the other. No. 23 gauge sheet is almost exactly 0.025 in. thick provided it is flat, so this will serve. Alternatively you can use a couple of No. 72 drills if you have them – 0.6 mm. Before going on to the drilling, mark out for the $\frac{3}{32}$ in. radii.

Use an eyeglass in drill-popping the positions with your spear-point, making a small indent first, and correcting it as you deepen the hole. Hold the work in a drill vice and drill all the holes in the face. 0.7 mm will serve instead of No. 70. The No. 57 hole in the end is offset from the

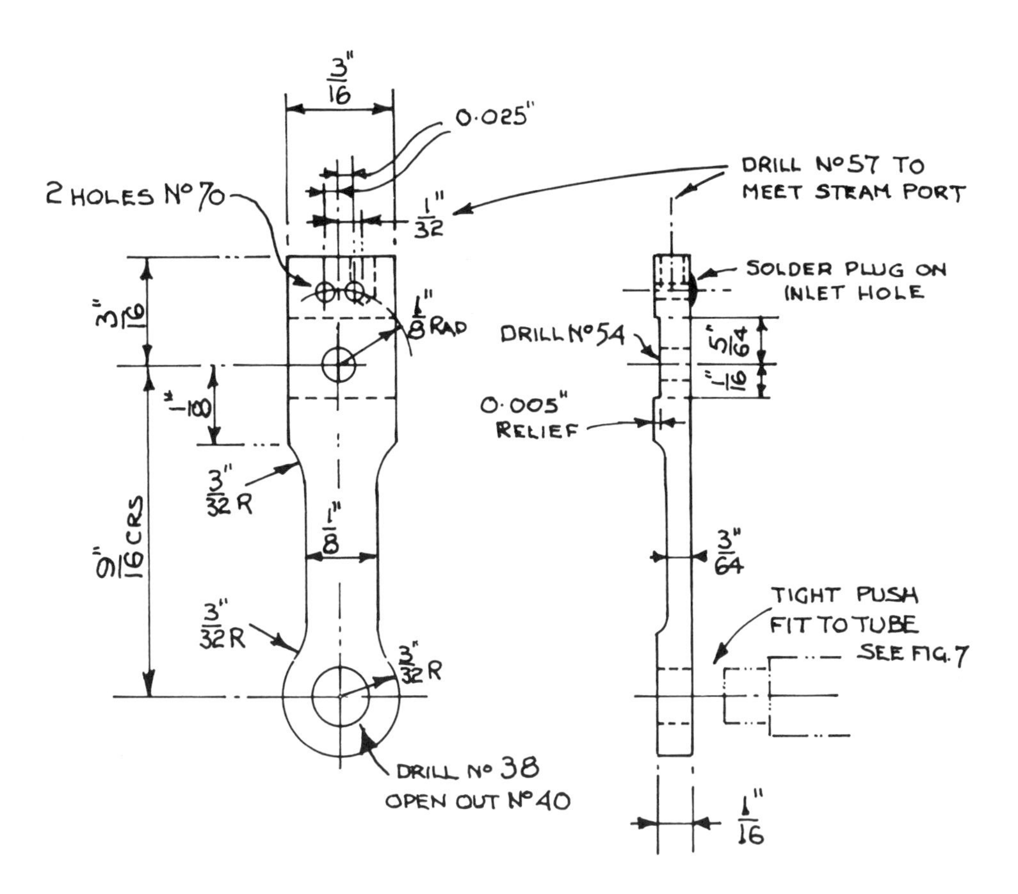

Fig 5–6 Standard

centreline of the port – actually at $\frac{1}{32}$ in. from the work centreline, but this isn't critical. The normal marking-out methods may be used. You can now form the shape and the recesses in the face. That marked 0.005 in. need only be a gesture of a recess, to concentrate the sealing pressure round the ports. You can now polish the faces *except* the portface itself – that on the left as you see the side view in the drawing. That face must be flattened on a glass plate as you did the cylinder face. Finally, put a small blob of solder on the inlet port back, but don't let it run into the port and block the passage.

Bearing Bracket. Fig 5–7

The bearing tube is made from $\frac{1}{8}$ in. dia. brass rod. Part off a piece $\frac{25}{32}$ in. long, having faced one end, and chuck this with say $\frac{3}{16}$ in. projecting. Centre and drill No. 47 for $\frac{5}{8}$ in. deep. Withdraw to clear chips frequently, to avoid running out of true. Follow this with a No. 51 drill right through. Now turn down the end to 0.098 in. dia. offering it up to the lower hole in the bearing standard as you do so. Again, with Loctite the need for an exact fit is less these days, but a tight push – not force – fit is best. Note that this section will be only 10 thou. thick for the $\frac{3}{32}$ in. length. Drill for the oilhole and remove the burr inside by putting the No. 47 drill down with your fingers.

The bracket itself is a case of "bend and try". You can drill the three No. 49 holes before bending if desired, and there is no reason why you shouldn't soften the 24 gauge brass if it offers any difficulties in getting the relatively sharp bends. The $\frac{11}{16}$ in. dimension isn't all that critical so long as there will be at least 15 thou. of the tube projecting at the larger end. It is more important that the curve bed well to the boiler shell, as everything is so light that you can't "pull" it in shape with the screws. Set out the $\frac{5}{32}$ in. dimension and drill the holes for the tube with a block of wood between the ears to avoid distortion, and then offer up the tube to see if it is square to the eye. Insert the 10 BA screw and tack it in place with solder, then fit the tube and solder that also, the solder being on the insides of the bracket.

Now offer the crankshaft to the tube. It should spin very freely. If it is stiff at the large end, try a 2mm drill in the hole, if at the smaller, try 1.75mm. But make sure there are no burrs first. If this fails check the shaft with your micrometer, and make sure the drills will give about 1 thou. clearance; if this is correct, then something has bent, and though enlarging the holes may make it "go" it won't be satisfactory if you need more than one size larger drill at either end, enlarging at the No. 51 end for preference. So, make a new one!

Attach the bracket to the boiler with the single screw, and mark through the other holes, after which drill the boiler No. 49. You may have a job holding the nuts, even with a box spanner. A little grease in the cavity of the box will help. Make sure, of course, that all is square before drilling, and if it looks cockeyed afterwards, draw or enlarge the holes to correct this.

Trial assembly.

With the bracket on the boiler, fit the standard to it and set it vertical by eye or by square; I find the former more reliable! Press it home till about 0.005 in. of the bearing tube is projecting from the face of the standard – just enough to give a bearing surface to the crank web. Insert the crank with a drop of oil and screw on the flywheel till there is just a trace of clearance sideways – again, 5 thou. is about right. If this thread is slack, then you must use Loctite screwlock – nothing stronger, as it may have to come off again. If there is excessive clearance when screwed home, use a 10 BA washer. Put in a drop of oil – Tellus 11 or Vitrea 25 as used on your lathe, not the "3-in-one" sort, though sewing machine oil will do at a pinch if all your cans are full of motor oil. A drop of oil on the port face, too, then fit the piston to the cylinder and the cylinder to the standard with the crankpin through the connecting rod hole. Fit the spring, and one nut to hold it there very lightly.

All should rotate freely, with breathing noises from the ports. If it doesn't, seek out and remedy the fault. If the cylinder seems to bear on one face and not the other, then the pivot pin is not square, but if this closes with say one turn of the spring nut it may be the ends of the spring are awry. If the faces part and close as the crank rotates, then the crankpin is not square to the web. You may be able to correct this, but if not, broach the crankpin bearing *very lightly* with a taper reamer. However, if anything is badly out it is doubtful if you will be able to correct by bending things as you might with a larger engine. The parts are too fragile. It is worth a try, but I would myself make a new part – she runs so well when right that it is worth the trouble.

You can now file down the end of a

Bearing Bracket Fig 5-7

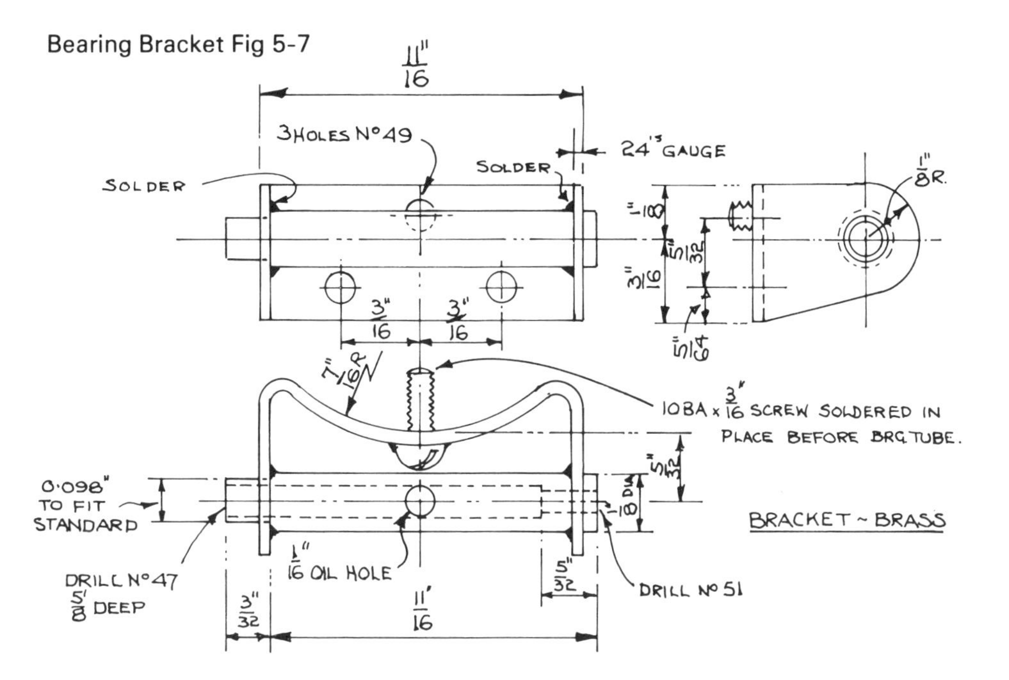

piece of $\frac{1}{16}$ in. pipe and hold it in the entry port whilst applying air, and I think you will be surprised at the way she goes. A bit stiff at first, but soon picks up. One point is worth mentioning. With both metric drills in steps of 0.05mm, and "number" drills available, it is possible to ease clearances on bearings etc. by very small steps indeed; No. 59 is one thou. bigger than No. 60, and 1.05mm is a bare half-thou. bigger again. So a little trial on these lines is well worth while. What should *NOT* be altered is the size of the steam and exhaust ports – though a quick 'shufti' down the holes to see that they do line up is an obvious precaution!

Steam Pipe Fig. 5–8

With the engine in its trial erection you can bend a piece of copper wire to a shapely curve from the steam hole in the boiler to the entry port of the standard, and use this to decide the length needed. The drawing shows the dimensions of mine, but each engine will differ slightly. The pipe itself is, however, another matter. I was feeling adventurous, and my own pipe is a piece of 18 gauge copper wire with a No. 68 hole drilled through. The wire was straightened first by gripping a few inches in the 3-jaw, taking hold of the other end with a chuck on the tailstock and applying traction. A piece about $1\frac{1}{8}$ in. long was cut off and faced both ends in the lathe – held in a collet – and after making a small centre hole the drill was applied. It was held in a brass collet in the tailstock with about $\frac{1}{8}$ in. protruding and as the drill penetrated this length was extended. Very frequent withdrawals for chip clearance were made, so that by the time full flute depth was reached only a few thou. increase in depth followed each withdrawal. The wire was then reversed and the hole drilled from the other end – to meet, much to my relief, in the middle.

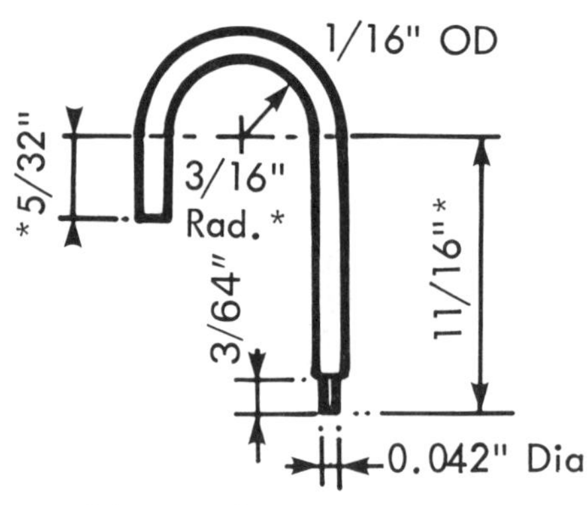

*Approximate see text

Steam Pipe Fig 5–8
Dimensions depend upon the type of tube used

I don't *recommend* this procedure, only mention it to show that it can be done. Some time later I acquired from that remarkable storehouse, Messrs Whiston's of New Mills, Stockport, a length of $\frac{3}{64}$ in. O.D. brass capillary tube; had I had this at the time I would have drilled with more confidence – the problem with deep hole drilling is always the risk that the drill may wander off centre and come out through the side; I imagine that oil-well drillers have a similar problem. So, if you do happen to have some of this tube, see how you get on. Don't forget – the tube must initially be straight, and you must clear chips every few thousandths of an inch of depth.

Failing this, we must use that which is commercially available – $\frac{1}{16}$ in. O.D. copper pipe, 24 or 26 gauge wall thickness. This is too large to solder to the standard, so the first step is to turn the end down as shown in the sketch, Fig. 5–8. The tube is then bent to your wire template, final adjustments being made on site. If you are satisfied with the whole at this stage, tighten up the three nuts holding the bracket to the boiler and, after removing the cylinder assembly, solder the pipe both to boiler and standard. Again, use the minimum of solder and make sure the flux is placed so that solder doesn't run all over.

Safety Valve Fig. 5–9

There is not much to this little fellow. The body is turned from brass rod; knurl it before doing any turning, it is easier that way. The end is faced and then drilled, after which it can be turned down to $\frac{5}{32}$ in. and screwed 40 t.p.i. with the tailstock dieholder. You may have to reverse the die to get the thread close to the shoulder – not normally recommended, but sometimes unavoidable. Make a little recess as shown on the drawing; I do this by hand turning, using a broken Swiss needle file with the end ground to a cutting edge. The little radius is formed with a roundnose tool before parting off, after which operation the burrs at the end of the hole must be removed. A very slight countersink will do no harm.

The problem with the ball is holding it. I have a pair of tweezers specially for holding such – I think they are jewellers "pearl tweezers" – and with these the ball can be inserted in the 3-jaw and fairly lightly gripped. You can either drill No. 54 and tap 10 BA as shown, or drill No. 51, drive the rod through and set a trace of solder on the protruding end. Either will do. Note that these balls are phosphor bronze, and do not drill easily; use a cutting oil and feed gently, clearing chips frequently. If you are lucky enough to get a nice curly chip coming from the drill, keep on feeding steadily, but as soon as this hesitates, stop at once and withdraw the drill to clear it.

The spring, like that for the cylinder pivot, may need some experiment. That shown is the same as mine, so try this to start with. Assemble the valve with the spring and screw on a brass nut. Borrow the kitchen scales and press down the end

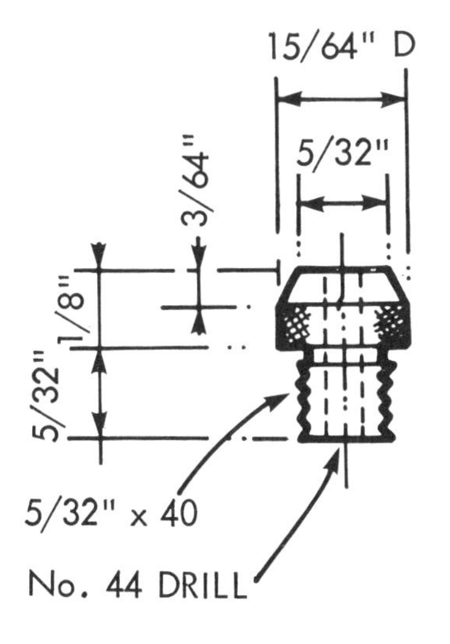

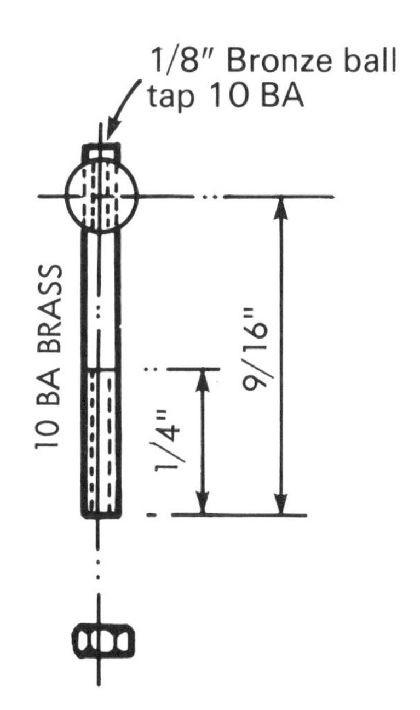

Brass nut. Solder
to spindle after
setting, reduce corners
to enter boiler bush.

Spring–30 G Bronze
5 free coils.
Free length – 5/16"

The Safety Valve Fig 5-9

of the rod on these, holding by the body. When the scales read $2\frac{3}{4}$ ounces and the ball just lifts, the valve will blow off at about 30 lb. sq. in. and pro rata. I set mine at this pressure. Don't solder the nut yet – wait till you have checked it under steam; very small changes in hole diameter make a big difference in the aiea at this scale. You need a little washer to seal the valve; I made mine from a bit of leather from a derelict pair of gloves, using a leather punch. Make two or three whilst you are at it – they are easily lost.

Lamp. Fig. 5–10

I think the drawing is more or less self-explanatory. The body can be made from tinplate or shimbrass – it could be as thin as 0.010 in. if of brass, but tinplate is easier to solder. There is no need to braze such a small lamp – it holds no more than 1 cc of spirit, which you can put out by slapping your hand over if need be. The two-part construction is necessary because if the spirit container is within the firebox the heat is too great and all evaporates before performing its office. The connecting pipe is $\frac{3}{32}$ in. copper tube – or brass – and if this is of the usual thick wall variety it is worth putting a $\frac{1}{16}$ in. drill through it, in the manner described above. The $\frac{5}{32}$ in. square bar under the reservoir is to support the lamp with the tube horizontal when the burner head is in the firebox. If you have departed from the dimensions of the base of the boiler you must adjust this support accordingly.

The firehead can be made from a discharged 0.22 in. cartridge, cut down, but make sure it *has* been discharged – there should be a little indentation at the rim somewhere; it is not enough that the shell be empty – the detonator in the rim will incommode you more than a little if it goes off whilst drilling; even worse if it doesn't go off till the lamp is lit! The wick can be a little bit from a "Kelly" lamp, a few shreds of asbestos string or even a bit of pyjama cord. I find, though, it needs renewing almost every time the engine is run, so have a little in hand.

Now, the lamp as shown will permit being refilled once or perhaps twice for one filling of the boiler. If you make the reservoir $\frac{11}{16}$ in. dia., the spirit/water heat ratio is about unity. (Don't make it any taller – the ratio between height of wick and height of firebox is a bit critical) However, if you do enlarge this it looks a bit out of proportion. I use a little syringe to fill the lamp, and the boiler also; one of the "disposable" hypodermics, 5 ml size, which gives a fair measure for both spirit and water. However, once the lamp is

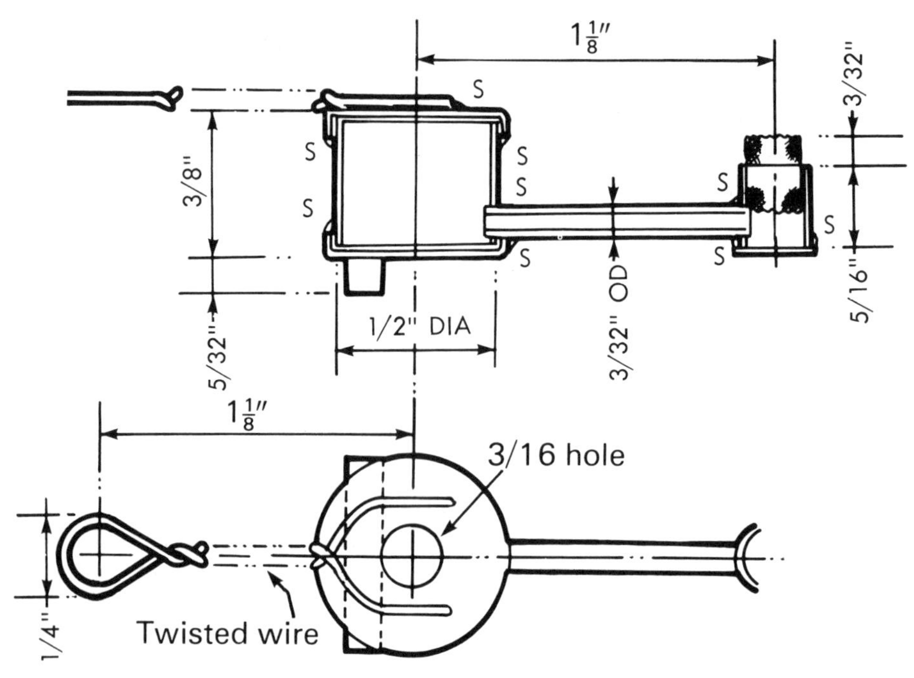

The Spirit Lamp Fig 5-10

made, fill it with spirit but without wick, and observe carefully for leaks. Resolder any that appear.

Running

Dismantle the "works" and check that the portfaces have bedded more or less. Oil all round – only a tiny drop on piston and portface, as excess of oil will set up a surface tension in the small ports that will resist full steam pressure. Don't forget a tiny drop on the crankpin, and one or two drops of thin oil on the main bearing tube. Reassemble, with, at this stage, the lightest pressure on the pivot spring. Fill with water from the kettle – as hot as the syringe will stand. 5 cc is about right, but if you use only three for the first attempt it will save time. Fill the lamp, insert, light and wait. After a while bubbles of water will emerge from the portface, condensation from the steam pipe. Rotate the engine slowly, and water should emerge from the exhaust – condensation from the cylinder. After a surprisingly short time she should warm through, and with a little hesitation, start to rotate.

If the safety valve blows at this stage, the spring is set too weak. If, when the engine is running, steam persistently escapes between the portfaces, either the faces are not true to each other, or the cylinder spring is set too weak. A little fiddling will set these to rights, but it may take a few runs before the engine really gets into good heart, with both piston and portface bedded down. Now with the lamp at full fire and the engine running, stop the wheel, with the cylinder inclined to the exhaust port. After a short while the safety valve should blow. If it doesn't, it is set too strongly. Once you have adjusted this valve, solder the nut in place. (You may have to file off surplus solder to get it through the bush afterwards, as there is very little clearance.) Fit a locknut to the pivot pin, taking care not to upset the tension.

That is all. You can now Brasso the whole, and, I suggest, make a little mount for it when not in use. Turn a nice plinth out of hard wood, dye it black and polish with beeswax. To get a glass cover, seek out a convenient chemical laboratory – the local Technical College or even the Path. lab at the Hospital, – and ask them to give you the name of a firm which makes large test-tubes. These can be had up to about 50 mm diameter and if you get the ordinary glass, not pyrex, are easily cut down to length. A little groove in the base to accept this cover, and a recess to hold the engine in place, and there you are; a true "glass case" model which can, when desired, run up to 3000 r.p.m. or more!

Tailpiece

As a novelty, try making an engine just like this but without a boiler, and with the crankshaft about half the length. Arrange the bearing bracket with a tie-pin, and connect the cylinder to a rubber bulb in your pocket. The decorative pin is certain to attract attention, and a squeeze on the bulb – you will be kept busy all night!

NOTES

NOTES

NOTES

NOTES